Integrated Science

Volume 22

The **Integrated Science** Series aims to publish the most relevant and novel research in all areas of Formal Sciences, Physical and Chemical Sciences, Biological Sciences, Medical Sciences, and Social Sciences. We are especially focused on the research involving the integration of two of more academic fields offering an innovative view, which is one of the main focuses of Universal Scientific Education and Research Network (USERN), science without borders.

Integrated Science is committed to upholding the integrity of the scientific record and will follow the Committee on Publication Ethics (COPE) guidelines on how to deal with potential acts of misconduct and correcting the literature.

Barbara W. K. Son
Editor

The Landscape of Global Health Inequity

Springer

Editor
Barbara W. K. Son
Anaheim University
Anaheim, CA, USA

ISSN 2662-9461 ISSN 2662-947X (electronic)
Integrated Science
ISBN 978-3-031-60501-7 ISBN 978-3-031-60502-4 (eBook)
https://doi.org/10.1007/978-3-031-60502-4

This Springer imprint is published by the registered company Springer Nature Switzerland AG
The registered company address is: Gewerbestrasse 11, 6330 Cham, Switzerland

I dedicate this book to my late mother, Margaret Bong S. Kim, who imparted invaluable values in life. Emphasizing wholesome health as paramount to everyone, this book endeavors to promote it.

Preface

The state of global health has reached a critical juncture, where disparities between resource-rich and resource-poor nations have become increasingly apparent. Despite significant strides in the containment of diseases like HIV/AIDS in countries such as the USA, many regions, particularly in Africa, continue to grapple with the devastating impact of infectious diseases. Furthermore, the global rise in cancer cases has accentuated the glaring disparities in access to cancer treatments, with limited options available in African nations.

The emergence of the COVID-19 pandemic exacerbated these inequalities, exposing the vulnerability of medical resource-poor nations in the face of global health crises. In this context, it becomes imperative to closely examine the lessons that the pandemic has taught us about addressing health disparities in regions with limited medical resources.

As we delve into the multifaceted issue of global health disparities, we must confront a range of obstacles, from cultural and socioeconomic barriers to stark differences in healthcare infrastructure. The stark reality is that the gap in healthcare access and outcomes between resource-rich and resource-poor countries is widening. It is a challenge that requires global, multi-level efforts to ensure that essential treatments and resources are accessible to all, regardless of their geographical location or economic status.

This Preface serves as an introduction to the complex web of issues surrounding global health equity. It is a call to action, highlighting the urgent need for concrete resolutions that will bridge the divide between nations. Access to health resources and the development of robust healthcare infrastructure should be at the forefront of our efforts, especially in developing countries, where the need is most acute. Clean water and adequate nutrition, fundamental to human well-being, remain elusive for far too many people in Africa and other resource-poor regions.

To achieve a comprehensive, integrated, sustainable, and equitable healthcare system, we must confront these challenges head-on. The path to reducing global health inequity is fraught with difficulties, but it is a path we must tread with determination, empathy, and unwavering commitment. Only by acknowledging the scale of the problem and by taking concrete steps toward solutions can we hope to make progress toward a more just and equitable global health landscape. This book entitled *The Landscape of Global Health Disparity* explores these challenges and aims to contribute to the discourse on creating a healthier world for all.

Los Angeles, USA
March 2024

Barbara W. K. Son, Ph.D.

Contents

Introduction to the Landscape of Global Health Inequity

1

Barbara W. K. Son

Abstract

The global stage is marked by a paradox—remarkable medical science and technology progress coexists with glaring disparities in health outcomes and access to care. While advancements in healthcare have led to the containment of once-devastating diseases in resource-rich nations, the burden of illness and mortality persists disproportionately in resource-poor regions. These inequities, starkly evident in areas like Africa, underscore the urgent need for a closer examination of global health disparities. *The Landscape of Global Health Inequity* is a journey into the heart of these disparities. In this introductory chapter, we set the stage for exploring the multifaceted challenges that define the global health landscape. These disparities are not isolated issues but an intricate tapestry woven from threads of unequal access to treatments and resources, limited clean water availability, inadequate nutrition, deeply entrenched socioeconomic and cultural barriers, the relentless persistence of infectious diseases, and fragile healthcare infrastructures. This book is not a mere exposé of the multifaceted challenges surrounding global health disparities. It is a call to action, an invitation to engage, question, learn, and act.

Keywords

Global health inequity • Gender • Minority • Pandemic • Infodemic • Global epidemics • Vaccination • Mental health • Clean water • Nutrition • Ecosystem

B. W. K. Son (✉)
Akio Morita School of Business, Anaheim University, Anaheim, USA
e-mail: bson@anaheim.edu

B. W. K. Son (ed.), *The Landscape of Global Health Inequity*, Integrated Science 22,
https://doi.org/10.1007/978-3-031-60502-4_1

1 Introduction

Global healthcare disparities result from various underlying factors. In most cases, these are regional disparities such as between the northern developed countries versus southern developing countries. However, in some cases, they are between groups within a country such as urban versus rural, or rich majority versus poor minorities. In some of the cases, these health concerns center around communicable diseases which like some of the recent pandemics, could quickly spread across the globe in a very short time. Hence, making this book a call for a global concern rather than simply an issue of the developing world or poor minorities.

1.1 General Observations

Global healthcare disparities result from various underlying factors. Barriers to healthcare equity, including socioeconomic, cultural, racial, ethnic, gender, and geographical barriers, contribute to this issue (Chap. 2). In low- and middle-income countries, non-communicable diseases like cancers surpass infectious diseases in causing deaths (Chap. 4). Africa is disproportionately affected by the HIV/AIDS epidemic, and while the disease is manageable with lifetime medication in high-income countries, African nations encounter challenges in acquiring and disseminating these drugs, contributing to global disparities in HIV/AIDS (Chap. 3). Furthermore, the COVID-19 pandemic has intensified the longstanding problem of medical inequalities in Africa (Chap. 5). The distribution of COVID-19 vaccine doses has underscored significant global inequities in procurement and allocation (Chaps. 6 and 7).

African nations grapple with a scarcity of fundamental necessities like clean water and sufficient nutrition, exerting a substantial impact on the health burden within underprivileged communities. Chapter 8 underscores the growing demand for clean water driven by population growth and urbanization in Africa. The manifestation of health disparities is evident in the prevalence of nutrition-related chronic diseases, disproportionately affecting disadvantaged populations, as discussed in Chap. 9. Chapter 10 highlights the significance of integrated big data and visual analytics in comprehending health and nutrition disparities.

Gender disparities in healthcare are thoroughly investigated in Chaps. 11 and 12. In Chap. 11, the focus is on examining socioeconomic and cultural obstacles faced by low-income women in Kenya. Chapter 12 delves into issues related to unmet contraception needs, gender-based violence, and intimate partner violence, particularly among adolescent women. Moving on to Chap. 13, the discussion centers on healthcare disparities affecting ethnic minorities. Specifically, it highlights the challenges of limited healthcare access encountered by tribal minorities in Indochina. These minorities, situated in Southeastern countries like Vietnam, Laos, Cambodia, and Myanmar, grapple with healthcare inequities due to factors such as insufficient infrastructure, remote locations, cultural barriers, and discrimination.

The multifaceted nature of these disparities involves elements like socioeconomic status and geographical constraints.

The escalation of global mental health disparities demands increased attention. Factors like stigma and limited access to mental health treatments contribute to these challenges (Chap. 14). Notably, mental health disparities are particularly pronounced within the Latino community. Specific obstacles to mental health services for Latinos are deeply rooted in cultural aspects, including the stigma surrounding mental health treatment and a lack of culturally competent resources for mental health services (Chap. 15). Finally, in Chap. 16, the exploration shifts to the One Health approach. This chapter delves into the historical origins of the approach, the ethical considerations associated with it, and the crucial importance of promoting equity and justice within the healthcare system.

1.2 Organization of the Book

This book is organized into four parts, the introduction; global disparities in treatments and resources; striving for health equity, challenges, and resolutions; and the conclusion. The book covers 16 chapters in total. Under global disparities in treatments and resources, the book is divided into 2 sections including global epidemics and health inequity; global pandemics and health inequity. For striving for health equity, challenges, and resolutions, the sections covered include water and nutrition resources; health equity for women and minorities; global and minority mental health inequity and resources; and the One Health approach to combat global health inequity. Finally, the book concludes with closing remarks and a call to action.

2 Global Disparities in Treatments and Resources

2.1 Global Epidemics and Health Inequity

Chapter 3 highlights a significant issue regarding the global distribution of healthcare resources, particularly in the context of HIV/AIDS and vaccine production. The disparities in access to treatment and preventive measures between developed and developing countries, particularly in Sub-Saharan Africa, have been a longstanding concern. HIV/AIDS is effectively managed in developed countries where medications and vaccines are easily obtainable. In contrast, the HIV/AIDS epidemic is predominantly prevalent in low-income, resource-limited nations, particularly in Africa, where more than 70% of global HIV-related deaths are reported [1]. Achieving the goal of eliminating the global health disparity in HIV/AIDS by 2030 appears challenging, given the substantial funding required for vaccine distribution. Compounding this challenge is the fact that Africa contributes less than one percent to the production of vaccines [2].

The global surge in cancer cases has brought to light another aspect of disparities in treatments and resources worldwide. Cancer stands as one of the leading causes of global mortality, with the cancer burden anticipated to exceed 85% in sub-Saharan Africa by 2030. However, only a small 5% of the global resources dedicated to cancer are allocated to sub-Saharan African countries [3]. In Kenya, cancer ranks as the third leading cause of death. Chapter 4 delves into the barriers on both the demand and supply sides that contribute to disparities in cancer treatment within the country. A comprehensive understanding of the interconnected elements of cancer care forms the basis for targeted policy interventions aimed at mitigating discrepancies in cancer treatment. The health system in Kenya confronts various challenges, marked by under-equipped hospitals, a low doctor-to-patient ratio, insufficient skills among service providers, a lack of cancer diagnoses and treatment facilities, and a low rate of cancer screening. Overcoming inequities in access to and enhancing the quality of cancer screening encounters additional obstacles. These include limited access to affordable cancer treatments, low community awareness regarding the importance of screening, a lack of knowledge about common cancer symptoms, inappropriate utilization of human resources, limited research and data to support policy formulation, and the concentration of cancer services in urban areas [3, 4]. Addressing these multifaceted challenges is imperative for promoting more equitable and accessible cancer treatment in Kenya.

2.2 Global Pandemics and Health Inequity

COVID-19 has disproportionately impacted individuals with lower incomes on a global scale [5, 6]. Low-income African nations, facing constrained resources and healthcare infrastructures, found themselves inadequately prepared to address the challenges posed by the COVID-19 pandemic. Chapter 5 scrutinizes these difficulties and draws insights from the experiences of the pandemic. To mitigate the disparities in access to pandemic treatments, a comprehensive and integrated approach is essential, involving active participation from stakeholders in both the public and private sectors. Simultaneously, efforts should focus on fortifying public health surveillance systems and enhancing public health education. Furthermore, inclusive strategies need to account for cross-sectoral partnerships and consider socioeconomic and cultural barriers to health equity, as discussed in Chaps. 2 and 5.

Chapter 6 underscores a critical concern related to the global distribution of COVID-19 vaccines. A reliable and effective COVID-19 vaccine that ensures optimal immunity stands as a more definitive method to curb the pandemic [5]. The inequitable distribution of COVID-19 vaccines is estimated to result in approximately 1.3 million deaths globally [7]. Fostering fair access to essential resources like vaccines necessitates collaborative efforts at multiple levels, involving coordination between international and individual policy-making entities, as well as non-governmental and grassroots organizations (Chap. 6). Additionally, strategies for optimal pandemic control must take into account constraints related to both health and financial resources, as explored in Chap. 7. When dealing with limited control resources, priority should be given to the high-risk group with an elevated risk of infection. Conversely, in situations where control resources are more abundant, efforts should be directed toward the low-risk group (Chap. 7).

3 Striving for Health Equity: Challenges and Resolutions

3.1 Clean Water Access and Nutrition

In Africa, clean water inequity stems from a complex interplay of interconnected causes. Chapter 8 thoroughly investigates the fundamental elements that contribute to the uneven distribution of access to clean water across the continent. The factors identified in this exploration encompass a comprehensive range, including climate change, urbanization, population growth, gender discrimination, inadequate infrastructure, corruption, mismanagement, and a lack of accountability. Adequate governance plays a pivotal role in ensuring effective water resource management [8]. Women and girls predominantly shoulder the responsibility of collecting water for their families. The necessity to traverse long distances to access water facilities not only impacts girls' education but also exposes them to the risk of violence [9]. The lack of access to clean water gives rise to diverse health challenges, encompassing issues such as hygiene and sanitation, nutrient absorption, and the occurrence of diseases from contaminated food and water. The intricate challenges associated with clean water inequity in Africa create a multifaceted framework that demands a comprehensive examination and targeted interventions.

Chapter 9 highlights the discernible manifestation of health disparities, particularly in the prevalence of nutrition-related chronic diseases that disproportionately impact disadvantaged populations [10]. These disparities result from various underlying factors, including income, education, culture, gender, and geographical barriers, as well as limited access to resources, significantly amplifying the burden of disease within these communities [11]. In the global effort to mitigate health disparities, it is crucial to acknowledge and address these challenges. Reliable and accurate nutrition data becomes imperative for developing strategies that more effectively target nutrition-related health issues at both national and global levels. Harnessing the potential of big data and advancements in visual analytics can significantly improve the global nutrition surveillance system, allowing for a thorough examination of health and nutrition disparities. In Chap. 10, the discussion centers on visuals guided by four principles: Evidence, Efficiency, Emphasis, and Ethics. The integration of big data and advancements in visual analytics provides an in-depth understanding of the multifaceted challenges and opportunities [12]. Emphasizing the crucial roles of integrated big data and food and nutrition dashboards, it highlights their effectiveness in disseminating global health and nutrition data. By shedding light on health and nutrition disparities, these tools become powerful instruments for formulating effective strategies to address the underlying causes of health inequities.

3.2 Health Equity for Women and Minority

Chapters 11 and 12 delve into the examination of gender disparities in low-income countries. Gender-based income gaps and affordability issues create substantial barriers to healthcare in Kenya. As of 2022, only 26% of women in Kenya had health insurance, according to the Kenya National Bureau of Statistics [13]. Additionally, cultural and accessibility challenges further complicate both treatment and health outcomes. To address the health equity gap, the Kenyan government should address disparities in gender resource allocation, boost the recruitment and training of female healthcare workers, offer cultural sensitivity training and education, and enhance health literacy through mobile phone technology, as outlined in Chap. 11. Chapter 12 delves deeper into the gender health inequities faced by adolescent women. Various obstacles, including resource limitations, socio-demographic factors, and cultural barriers, hinder access to health and reproductive services. Resource challenges involve inadequate infrastructure, facility services, and education. Socio-demographic and cultural barriers encompass factors such as age, gender-based violence, patriarchal culture, discrimination, religious values, and the absence of integration of family planning programs into comprehensive reproductive and sexual health services [14, 15]. Therefore, it is crucial to adopt integrated strategies that prioritize cross-sector and cross-program collaboration, gender-specific services, and improve affordability and access to health and reproductive services for adolescent women, as emphasized by the UNFPA in 2022 [16].

In Chap. 13, an exploration of health equity disparities unfolds, focusing on tribal minorities in Indochina. The chapter scrutinizes factors contributing to specific health challenges faced by these minorities, including malnutrition, restricted access to healthcare services, inadequate healthcare infrastructure, geographical and linguistic obstacles, and cultural incompetency among healthcare workers [17–19]. The chapter underscores the importance of a multi-sectoral approach, advocating for partnerships between governments, non-governmental organizations, and tribal communities. It calls for the implementation of a comprehensive and inclusive healthcare system aimed at enhancing healthcare infrastructure, improving accessibility in tribal areas, and providing health education to address the unique health needs of tribal minorities.

3.3 Mental Health and Health Inequity

Mental health disparities, exacerbated by cultural barriers, limited access to mental health services, and a shortage of mental health professionals, are alarming concerns discussed in Chaps. 14 and 15. The stigma surrounding mental health concerns discourages individuals from seeking the necessary services, as highlighted in Chap. 14. Despite the prevalence of self-reported mental illness being similar for both Latinos in the U.S. (22%) and non-Latino Whites (23.9%), there exists a significant discrepancy in the utilization of mental health services. Only 36% of Latinos access these services compared to 52% of non-Latino Whites [20]. The shortage of bilingual mental health professionals contributes to difficulties in effective communication with healthcare providers [Chap. 15]. Addressing mental health needs and preventing and treating mental health issues requires a holistic and multidisciplinary approach, as emphasized in Chap. 14. This approach is crucial to overcoming the complex challenges associated with mental health, including cultural barriers and disparities in access to services.

3.4 Striving Toward Global Health Equity: Building a Sustainable and Equitable Ecosystem

Human health is intimately intertwined with the ecosystem. Degradation of the environment has resulted in the spread of infectious diseases among both humans and animals, disproportionately affecting marginalized communities lacking adequate resources (Chap. 16). The contamination of water sources and inadequate water and sewage treatment have contributed to the prevalence of waterborne and zoonotic diseases, particularly impacting low-income nations [21]. Consequently, disparities in healthcare are closely linked to insufficient environmental resources and limited access to veterinary services. Addressing the breadth of these inequalities is imperative to advance global health equity through initiatives like One Health, which prioritize fair access to healthcare and environmental well-being [22]. Therefore, it is crucial to adopt a comprehensive and interdisciplinary approach that fosters a sustainable and fair ecosystem [23].

4 Conclusion

Our examination delved into global health disparities across a spectrum of areas, spanning epidemics, pandemics, inequalities concerning women and minorities, disparities in medical and healthcare infrastructure, research and data, and access to environmental resources. These disparities persist due to a multifaceted interplay of unequal access to treatments and resources, exacerbated by socioeconomic, cultural, educational, and geographical barriers. Specifically, marginalized communities with low incomes encounter numerous deficiencies in critical resources like medical care, clean water, food, nutrition, and sanitation. However, there are promising strategies proposed to tackle these pressing issues.

As we navigate the intricate terrain of global health inequality, it is crucial to remember that the journey towards a fairer world may be arduous, yet it remains a noble pursuit demanding steadfast dedication and empathy. "Exploring the Global Health Disparity Landscape" extends an invitation to join us in this endeavor, to delve into the complexities of global health disparities, and to engage in discussions that will shape the future of healthcare worldwide. Together, let us strive for a world where health is not constrained by borders, and where every individual has the opportunity to flourish.

5 Core Messages

- The book presents compelling examples that underscore the notable health disparities within intricate and diverse contexts across various regions.
- It thoroughly examines the interconnected and multifaceted root causes that are prevalent in developing countries.
- The book presents crucial and comprehensive recommendations advocating for collaborative endeavors at global, national, and local levels to effectively tackle the identified issues.

References

1. Kharsany ABM, Karim QA (2016) HIV infection and AIDS in Sub-Saharan Africa: current status, challenges and opportunities. The Open AIDS J 10:34–48. https://doi.org/10.2174/1874613601610010034
2. Nweneka CV, Disu T (2022) The future of vaccine manufacturing in Africa. Foresight Africa. https://www.brookings.edu/articles/the-future-of-vaccine-manufacturing-in-africa/
3. Wambalaba F, Son B, Wambalaba A, Nyong'o D, and Nyong'o A (2019) Prevalence and capacity of cancer diagnostics and treatment: a demand and supply survey of health-care facilities in Kenya. Cancer Control 26(1). https://doi.org/10.1177/1073274819886930
4. Morgan C, Cira M, Karagu A, Asirwa FC, Duncan K (2018) The Kenya cancer research and control stakeholder program: evaluating a bilateral partnership to strengthen national cancer efforts. J Cancer Policy. 17:38–44. https://doi.org/10.1016/j.jcpo.2017.03.012
5. Son B (2023) A multipronged approach to combat covid-19: lessons from previous pandemics for future. Ch. 4. Integrated Science of Global Epidemics. Springer, pp73–92. ISBN: 978–3–031–17777–4
6. Okonkwo NE, Aguwa UT, Jang M, Barré IA, Page KR, Sullivan PS, Beyrer C, Baral S (2020) COVID-19 and the US response: accelerating health inequities. BMJ Evid Based Med 26(4):176–179. https://doi.org/10.1136/bmjebm-2020-111426
7. Ledford H (2022) COVID vaccine hoarding might have cost more than a million lives. Nature.https://doi.org/10.1038/d41586-022-03529-3
8. Phillips SD (2019) Proximities of violence: civil order beyond governance institutions. Int Stud Quart 63:680–691. https://doi.org/10.1093/isq/sqz036
9. UNICEF (2022) Africa to drastically accelerate progress on water, sanitation and hygiene—report. [Press release]. Dakar, Senegal, UNICEF Senegal. https://www.unicef.org/senegal/en/press-releases/africa-drastically-accelerate-progress-water-sanitation-and-hygiene-report
10. Satia JA (2009) Diet-related disparities: understanding the problem and accelerating solutions. J Am Diet Assoc 109(4):610–615. https://doi.org/10.1016/j.jada.2008.12.019
11. Micha R, Shulkin ML, Peñalvo JL, Khatibzadeh S, Singh GM, Rao M, Fahimi S, Powles J, Mozaffarian D (2017) Etiologic effects and optimal intakes of foods and nutrients for risk of cardiovascular diseases and diabetes: systematic reviews and meta-analyses from the Nutrition

and Chronic Diseases Expert Group (NutriCoDE). PLoS ONE 12(4):e0175149. https://doi.org/10.1371/journal.pone.0175149
12. Zhou B, Liang S, Monahan KM, Singh GM, Simpson RB, Reedy J, Zhang J, DeVane A, Cruz MS, Marshak A (2022) Food and nutrition systems dashboards: a systematic review. Adv Nutr 13(3):748–757
13. Kenya National Bureau of Statistics (2022) Kenya demographic and health survey. Kenya National Bureau of Statistics, Nairobi
14. Asmamaw DB, Negash WB (2023) Unmet need for family planning and associated factors among adolescent girls and young women in Ethiopia: a multilevel analysis of Ethiopian demographic and health survey. Contracept Reprod Med 8(1):1–11. https://doi.org/10.1186/s40834-022-00211-x
15. Rizvi F, Williams J, Bowe S, Hoban E (2020) Factors influencing unmet need for contraception amongst adolescent girls and women in Cambodia. Peer J 8:e10065. https://doi.org/10.7717/peerj.10065
16. UNFPA (2022) UNFPA strategy for family planning, 2022-2030: expanding choices—ensuring rights in a diverse and changing world. Laser Focus World 39(11):60
17. Gilmour B, Alene KA, Clements A (2022) The prevalence of tuberculosis and malaria in minority indigenous populations of South- East Asia and the Western Pacific Region: a systematic review and meta-analysis. Pathog Glob Health 116(4):201–219
18. Akter S, Davies K, Rich JL, Inder KJ (2020) Barriers to accessing maternal health care services in the Chittagong Hill Tracts, Bangladesh: a qualitative descriptive study of Indigenous women's experiences. PLoS One 15(8):e0237002
19. Roesler A, Smithers LG, Winichagoon P, Wangpakapattanawong P, Moore V (2018) Health workers' and villagers' perceptions of young child health, growth monitoring, and the role of the health system in remote Thailand. Food Nutr Bull 39(4):536–548
20. Substance Abuse and Mental Health Services Administration (2022) Key substance use and mental health indicators in the United States: results from the 2021 National Survey on Drug Use and Health (HHS Publication No. PEP22–07–01–005, NSDUH Series H-57). Center for Behavioral Health Statistics and Quality, Substance Abuse and Mental Health Services Administration. https://www.samhsa.gov/data/report/2021-nsduh-annual-national-report
21. Mutono N, Wright J, Mutembei H, Muema J, Thomas M, Mutunga M, Thumbi SM (2020) The nexus between improved water supply and water-borne diseases in urban areas in Africa: a scoping review protocol. AAS Open Res 8(3):12. https://doi.org/10.12688/aasopenres
22. Riadhussyah M, Jannah AW, Basir SM, Subarkah AR (2023) Forming international instrument through one health approach for health justice. J IUS Kajian Hukum dan Keadilan 11(3):493–511. https://doi.org/10.29303/ius.v11i3.1246
23. Adisasmito WB, Almuhairi S (2023) One health action for health security and equity 401(10376):530–533. https://doi.org/10.1016/S0140-6736(23)00086-7

Infectious Diseases and Global Health Inequity

2

Nadim Sharif and Shuvra Kanti Dey

Abstract

Infectious diseases are defined as a group of illnesses requiring the involvement of pathogenic microorganisms including, bacteria, viruses, parasites, fungi, and protozoa, and can be transmitted from one person to another person via direct or indirect contact. Infectious diseases and health inequity are closely related. They both have significant impact on local and global public health. Infectious diseases like COVID-19, tuberculosis, influenza, polio, malaria, etc. can have a significant influence on health inequity in a society. On the other hand, health inequity can influence the prevalence of many infectious diseases in a society. Health inequity can be defined as disparities in health outcomes or access to healthcare services between different populations or social groups. Apart from infectious diseases, various factors including socioeconomic status, race, ethnicity, gender, education, and geographical location can influence these health disparities. Recently, the onset and spread of COVID-19 led to one of the largest pandemics in human history. The disease itself and the preventive measures have influenced health disparities among people from different societies and regions globally. However, there is a huge gap in studies focusing on the impact of COVID-19 on health inequity. This chapter explored the impact of infectious diseases like COVID-19 on health inequity and vice-versa. Reports of COVID-19 pandemics and other infectious diseases concisely documented that inequality, poverty, and socio-economic determinants of health influence the transmission of infectious diseases. International policy-making bodies including WHO to individual countries, must adapt policies on social determinants of health and health inequalities. The integrated approaches must include limiting the transmission of infectious diseases by improving the surveillance system, revising the definition of risk factors for disease, and assessing the impact of

N. Sharif (✉) · S. K. Dey
Department of Microbiology, Jahangirnagar University, Savar, Dhaka 1342, Bangladesh
e-mail: nadim@juniv.edu

B. W. K. Son (ed.), *The Landscape of Global Health Inequity*, Integrated Science 22,
https://doi.org/10.1007/978-3-031-60502-4_2

alternative policies on prevention. Alternatively, policies should be made to reduce health disparities that can contribute to reduced burden of infectious diseases.

Keyword

Health inequities • Infectious diseases • Pandemic • LMICs

1 Introduction

Global health inequity is one of the major challenges both for developed and developing countries [1, 2]. According to the World Health Organization, "*Health inequities are systematic differences in the health status of different population groups*" These inequities can have a significant impact on the social and economic status of individuals and societies. Different factors including gender, education, ethnicity, employment status, race, and income level, can significantly influence the health of an individual [1–3]. Different disparities in the health of people in various countries including low-income, middle-income, and high-income exist among different socio-economic groups [1, 3–6]. According to the socio-economic position and other factors, the health status varies from person to person. Health inequities vary according to the socio-economic conditions of different places such as birthplace, growing areas, living, working, and aging areas.

WHO defines health as, "*Health is a state of complete physical, mental and social well-being and not merely the absence of disease or infirmity*". The quality and distribution of health among people from different socio-economic statuses is a predictor of development [1, 6–9]. Health is one of the fundamental human rights and interconnected with all other basic rights. Thus, degradation and any inequity in health are a violation of human rights [3, 10].

Health inequity and health inequality are different terms used to describe health access and status in a community [1, 3, 11–13]. Health inequality arises when the health of two different demographic groups varies despite access to similar health facilities. For example, if one group dies earlier than another due to a non-remediable factor, it is defined as health inequality. Alternatively, if a population is deprived of equal rights to access preventive measures, medications, and health-care facilities, the situation is called health inequity. These inequities significantly influence health outcomes in a population [2, 13–15]. Global health equity is a crucial point for achieving the Millennium Development Goals [1, 3, 15, 16].

Infectious diseases are caused by different pathogenic microorganisms. These infectious pathogens are viruses, bacteria, fungi, and parasites [2, 15–17]. They have been present with us since the first day of our journey on this planet. Pathogenic microorganisms can be transmitted to us both via direct and indirect ways. Infectious disease is a major cause of sickness and death worldwide and a significant predictor of health [17–22]. Every year an estimated 1.5–2 billion new cases of infectious diseases are reported and 10–14 million people die of them globally [18, 20, 23]. The emergence of an endemic or pandemic affects both local and global health within a very short period. Recently, the emergence and global transmission of COVID-19, AMR pathogens, H7N9 influenza, MERS-CoV, and dengue have created a health crisis [17–24].

Nowadays outbreaks of infectious disease can be transmitted faster worldwide, due to faster travel and communication across countries and continents. Historical backgrounds of cholera, malaria influenza, and recent pandemics distinctly documented that social factors, poverty, economic status, and inequality contributed to the transmission of infectious diseases. Further, the existing health inequalities can contribute to unequal outcomes of morbidity, mortality, and health status globally [18–23]. Till now, there exists a significant lack of studies on the impact of infectious diseases on health inequity and vice versa. In this chapter, we explored the impact of infectious diseases on global health inequity and vice-versa.

2 Infectious Diseases Affecting the Global Health

About one-third of global deaths occur due to infectious diseases every year. Infectious diseases are one of the major causes of human suffering in terms of both morbidity and mortality [18]. The transmission of infectious diseases was fueled by different footsteps of our civilization. Such as zoonotic and parasitic diseases have become more prevalent after the domestication of birds and animals. Larger pandemics including smallpox, plague, cholera, and influenza have occurred since the historical period [21–24]. Further, chronic infectious diseases including AIDS, tuberculosis, and syphilis are also negatively influencing global health. Recent pandemics of COVID-19, AMR, and dengue have impacted billions of lives worldwide.

According to WHO, about 500 million cases of dengue, 300–500 million cases of malaria, 350 million cases of sexually transmitted diseases, 250 million cases of AMR, 200 million cases of diarrheal illness, 50 million cases of HIV/AIDS, 20 million cases of tuberculosis, and 3–5 million cases of cholera occur annually [1, 3]. Recently, the COVID-19 pandemic has infected over 0.7 billion people. Infectious diseases have a significant impact on the lives of people [25]. Apart from mortality, people suffer from losses of economic status, social status, and chronic disabilities due to infection by pathogens [25–28]. Among the infectious diseases, lower respiratory tract infection causes the highest death followed by HIV/AIDS, diarrheal diseases, tuberculosis, malaria, and measles, respectively (Table 1). Though this data represents a consistent health burden of infectious diseases over the last hundred years, the sudden onset of a pandemic like COVID-19 has a more severe impact on people worldwide.

Table 1 An estimated cause of mortality per-year by infectious diseases globally [11]

Cause of death	Deaths (in millions)	Percentage of all deaths (%)	Endemic regions/ prevalent
Lower respiratory infections	3.9	6.9	Middle-East, Asia
HIV/AIDS	2.8	4.9	Asia, North America, Africa
Diarrheal diseases	1.8	3.2	Asia, Africa
Tuberculosis (TB)	1.6	2.7	Asia
Malaria	1.3	2.2	Asia, Africa
Measles	0.6	1.1	Africa, Asia
Pertussis	0.29	0.5	Asia
Tetanus	0.21	0.4	Africa
Meningitis	0.17	0.3	Africa, North America
Syphilis	0.16	0.3	Africa
Hepatitis B	0.10	0.2	Asia
Tropical diseases	0.13	0.2	Asia
All infectious diseases	14.7	25.9	Asia, Africa

2.1 Impact of COVID-19 Pandemic on Health

The COVID-19 pandemic has become one of the top public health issues since the first report in December 2019. COVID-19 has transmitted faster across the border and affected people of all aspects worldwide [28, 29]. Nearly 0.7 billion cases and 6 974 186 fatalities of COVID-19 have been reported globally [30]. During the pandemic, every aspect of our lives has been affected severely. The tremendous pressure of COVID-19 on the healthcare system and healthcare professionals disrupted normal activities globally. Before the use of vaccines, several preventive measures affected the global economy, transportation, communication, and normal life activities significantly. The pandemic has affected both the health system and services. Lockdown and quarantine of infected individuals and regions affected every aspect of human life [31–35].

During the pandemic, about 90% of normal health services were disrupted. As economic conditions degraded, the capability of people was reduced to get access to health facilities [31–34]. Treatment of COVID-19 patients with critical conditions created an economic burden. People from all over the world were infected by COVID-19. Due to the poor economic conditions of many patients, they could not afford the cost of treatment for COVID-19 in severe conditions. The majority of the deaths associated with COVID-19 could be minimized if proper ICU facilities and oxygen supply were provided to the patients. The incident of inequal health access according to economic status came to light during the pandemic. People with higher economic status got ICU support and other emergency facilities more easily than people with lower economic status [32–37].

During the pandemic, many people lost their jobs and the daily workers also became jobless. So, people from both developed and developing countries suffered from severe economic losses. At the same time, they were infected by COVID-19 and required healthcare facilities. So, the economic and health crisis created by the pandemic influenced and contributed to health inequity globally [36–39]. After the onset of COVID-19, healthcare professionals struggled due to the increased number of patients, lack of facilities dedicated to COVID-19, increased workload and mental pressure, and lack of training and personal protective equipment [36, 37].

The pandemic affected both the patient care provider and service receiver concepts [37, 38]. Such as the professionals like nurses and ward boys became more vital during the pandemic. They felt more confident and took their responsibilities as vital to providing care to the patients. However, in resource-limited settings the disparities for providing services were severe. With limited resources and constricted economic conditions, healthcare professionals in developing countries felt helpless during the start of the pandemic.

2.2 COVID-19 and Economic Crisis

The COVID-19 pandemic has created longer economic impacts than our imagination. The world has faced one of the major recessions during the pandemic [35–40]. All forms of business have decreased during the COVID-19 lockdowns in every part of the world. As a result, the stock market crashed and created a financial crisis [35–38]. A cycle of events associated with economic stress such as shortage of liquid money in the banks, inflation surge, price hikes, shortages of products, and imbalanced pricing of medical and health products happened during 2020–2023 Shortage of food products and essential commodities affected people worldwide. Such economic imbalance and crisis had a far-reaching impact on the health of the majority of the people in developing and poor countries. The degradation of social and economic status during the pandemic created a significant health inequity among the majority of people worldwide [36, 37].

2.3 COVID-19 and Inequality

Lower and middle-class people were more likely to get COVID-19 and suffer from both social, economic, and health crises during the pandemic [40–43]. In developed countries, low-income neighborhoods and people with lower social status were disproportionately affected by COVID-19 and after COVID-19 complications. For example, uninsured people in developed countries like the United States, the UK, and Spain were deprived of rapid access to healthcare facilities. Billions of people lost their jobs and were bound to shut down businesses. As a result, they also lost their health insurance and other health accesses that were available to them before the pandemic. All these events worsened the existing health inequity in the developed countries and introduced severe inequity in the developing and poor countries [40–46].

3 Infectious Diseases and Inequality

Infectious disease outbreaks in any local area can become a global health issue within a short time as global interconnection and transportation have increased significantly [2–7]. Infectious diseases cannot be confined within borders or continents. We have experienced repeatedly global transmission of COVID-19, HIV/AIDS, influenza, tuberculosis, and monkeypox. The way infectious diseases affect our health system, society, economic status, and inequity, similarly social determinants of health, poverty, social and economic status, and inequity can influence the successful onset and persistence of a pandemic (Fig. 1) [8–13]. The COVID-19 pandemic has worsened the existing health inequalities [41–43]. Further, the disproportionate distribution of morbidity and mortality originated and sustained in lower- and middle-income countries (LMICs) for a long time. For example, during the COVID-19 vaccination campaign, developed countries got the privilege of getting vaccines and medicines. Health disparities are disproportionately distributed for the majority of the infectious diseases in developed and LMICs [40–45]. If we do not take steps until the developed countries are hit by infectious diseases, we will continue to see a heavy burden of global health disparities, morbidity, and mortality.

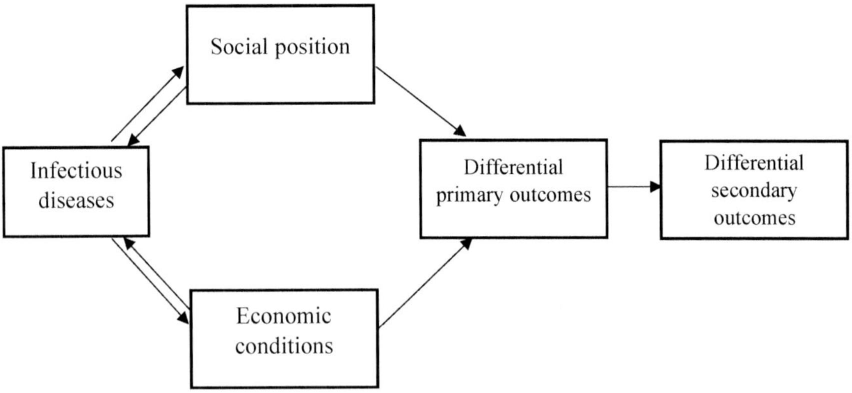

Fig. 1 A cyclic framework on the impact of infectious diseases on health outcomes. *Note* Differential primary outcomes included transmission, susceptibility, exposure and consequences and secondary outcomes included morbidity and mortality

4 Facts of Real World Health Inequity

Health inequities have a long-term impact on global health. Though individuals in LMICs are affected directly, the long-term effects are observable in developed countries. The burden of health inequities can be minimized by the mix of international and national policies and the integration of bodies.

4.1 Maternal and Child Mortality as an Indicator of Health Inequity

Both the maternal and child mortality rates are indicators of the health status of a population. Maternal mortality can depict the actual gaps between rich and poor, both between and within countries. About 99% of maternal deaths occur in the developing countries every year [11, 14, 16]. For example, women in a country like Chad have a higher risk of maternal death of 1 in 15, while women in a developed country like Sweden have a risk of maternal death of less than 1 in 10,000. Mortality rates among children due to diarrhea, malaria, pneumonia, and other infectious diseases are significantly higher in LMICs [7–11]. The death rate among children under five years is about 14 times higher in the countries in sub-Saharan Africa. Moreover, children residing in rural and poorer areas are still disproportionately affected [11].

4.2 Life Expectancy and Health Inequities Varies Between Cities and Countries

Life expectancy is one of the major indicators of health status. The average life expectancy is about 60 years in LMICs, while, it is around 80 years in the higher income countries. A child in Sierra Leone can have a life expectancy of 50 years. On the other hand, a child in Japan can have a life expectancy of 85 years [11]. In countries like the USA and UK, life expectancy varied between neighborhoods. In areas with residents of lower-income families, the health status is poorer than in areas with individuals having higher incomes.

4.3 Disease of Poverty

Several infectious diseases are already known as diseases of poor people globally. Tuberculosis is such a disease. About 95% of TB-related deaths are documented in the developing countries [8–11]. This morbidity and mortality affect mainly young children and adults in developing and poor countries. TB has a far-reaching impact on the economic and social status of the infected individuals and their families.

5 Health Equity for Health Security

Increased global vulnerability to infectious diseases has created new security challenges for global public health. Globalization, rapid transportation, and transnational communication have contributed to the mass transmission of infectious diseases like COVID-19, mpox, and dengue globally [1–4]. As a result, the management of infectious diseases has become a burning issue in both national and international security policies. Infectious diseases, epidemics, and pandemics have become alarming issues for global policymakers and local governments [46–49]. Both national and international authorities should work together to improve the detection, treatment, and prevention.

6 Conclusion

Infectious diseases are major contributors to health inequities globally. Individuals who are disadvantaged by health inequities suffer from worse health outcomes than those with access to better facilities. Infectious diseases and pandemics like COVID-19 have serious consequences on every aspect of our lives. The onset of the COVID-19 pandemic has worsened existing health inequities globally. To minimize the losses and long-term health inequities, the spread of infectious diseases should be controlled. Therefore, researchers, healthcare providers, and policymakers should work together to prevent the massive spread of infectious diseases and provide equal facilities to people of different economic and social statuses. All shall come forward to minimize the unequal distribution of morbidity and mortality worldwide.

Declaration of Competing Interest
The authors declare that they have no known competing financial interests or personal relationships that could have appeared to influence the work reported in this paper.

Funding
None.

Acknowledgements This research did not receive any specific grant from funding agencies in the public, commercial, or not-for-profit sectors.

Data Availability
All the necessary data have been made available in the Article and Supplementary appendix.

Author Contributions
Nadim Sharif: writing—original draft preparation (lead), methodology (lead), investigation (lead), conceptualization (lead), project administration (lead), writing—review and editing (equal), formal analysis (lead), data collection, data analysis, data interpretation (lead).

Shuvra Kanti Dey: literature search (lead), figures (lead), study design, data collection, data analysis, data interpretation (lead), supervision (lead).

References

1. Quinn SC, Kumar S (2014) Health inequalities and infectious disease epidemics: a challenge for global health security. Biosecur Bioterror 12(5):263–273
2. Farmer P (2016) Social inequalities and emerging infectious diseases. Understand Appl Med Anthropol 118–126
3. Global Report for Research on Infectious Diseases of Poverty (2012) WHO. http://whqlibdoc.who.int/publications/2012/9789241564489_eng.pdf?ua=1. Accessed on 17 Feb 2024
4. Closing the Gap in a Generation: Health Equity Through Action on the Social Determinants of Health. Geneva: World Health (2008). http://whqlibdoc.who.int/publications/2008/9789241563703_eng.pdf. Accessed on 17 Feb 2024
5. Blumenshine P, Reingold A, Egerter S, Mockenhaupt R, Braveman P, Marks J (2008) Pandemic influenza planning in the United States from a health disparities perspective. Emerg Infect Dis 14(5):709
6. Quinn SC, Kumar S, Freimuth VS, Musa D, Casteneda-Angarita N, Kidwell K (2011) Racial disparities in exposure, susceptibility, and access to health care in the US H1N1 influenza pandemic. Am J Public Health 101(2):285–293
7. World Health Organization (2024) Health inequality data repository. Health Inequality Monitor. https://www.who.int/data/inequality-monitor/health-inequality-data-repository. Accessed on 17 Feb 2024
8. Hosseinpoor AR, Bergen N, Kirkby K, Schlotheuber A (2023) Strengthening and expanding health inequality monitoring for the advancement of health equity: a review of WHO resources and contributions. Int J Equity Health 22(1):1–9
9. World Health Organization (2024) State of inequality: HIV, tuberculosis and malaria. World Health Organization, Geneva. Accessed on 17 Feb 2024
10. Global Fund to Fight AIDS, Tuberculosis and Malaria (2021) Fighting pandemics and building a healthier and more equitable world: global fund strategy (2023–2028). https://www.theglobalfund.org/en/strategy/. Accessed on 17 Feb 2024
11. World Health Organization (2023) Health inequality monitoring. Open WHO. https://openwho.org/channels/inequality-monitoring. Accessed on 17 Feb 2024
12. Kirkby K, Bergen N, Baptista A, Schlotheuber A, Hosseinpoor AR (2023) Data resource profile: world health organization health inequality data repository. Int J Epidemiol 52:e253–e262
13. Bambra C (2022) Pandemic inequalities: emerging infectious diseases and health equity. Int J Equity Health 21(1):6
14. Ayorinde A, Ghosh I, Ali I, Zahair I, Olarewaju O, Singh M, Meehan E, Anjorin SS, Rotheram S, Barr B, McCarthy N (2023) Health inequalities in infectious diseases: a systematic overview of reviews. BMJ Open 13(4):e067429
15. Mamelund SE, Dimka J (2021) Social inequalities in infectious diseases. Scandinavian J Public Health 49(7):675–680
16. Lim SS, Mokdad AH (2012) Socioeconomic inequalities and infectious disease burden. The Lancet 379(9821):1080–1081
17. Heywood AE, López-Vélez R (2019) Reducing infectious disease inequities among migrants. J Travel Med 26(2):tay131
18. Becker K, Hu Y, Biller-Andorno N (2006) Infectious diseases–a global challenge. Int J Med Microbiol 296(4–5):179–185
19. Baker RE, Mahmud AS, Miller IF, Rajeev M, Rasambainarivo F, Rice BL, Takahashi S, Tatem AJ, Wagner CE, Wang LF, Wesolowski A (2022) Infectious disease in an era of global change. Nat Rev Microbiol 20(4):193–205

20. Bhutta ZA, Sommerfeld J, Lassi ZS, Salam RA, Das JK (2014) Global burden, distribution, and interventions for infectious diseases of poverty. Infect Dis Poverty 3(1):1–7
21. Harris M, Reza JN (2012) Global report for research on infectious diseases of poverty. World Health Organization. Accessed on 17 Feb 2024
22. Nii-Trebi NI (2017) Emerging and neglected infectious diseases: insights, advances, and challenges. Biomed Res Int 245021:1–15
23. World Health Organization (2009) Global health risks: mortality and burden of disease attributable to selected major risks. World Health Organization. Accessed on 17 Feb 2024
24. Bloom DE, Cadarette D (2019) Infectious disease threats in the twenty-first century: strengthening the global response. Front Immunol 10:549
25. Sharif N, Alzahrani KJ, Ahmed SN, Khan A, Banjer HJ, Alzahrani FM, Parvez AK, Dey SK (2022) Genomic surveillance, evolution and global transmission of SARS-CoV-2 during 2019–2022. PLoS ONE 17(8):e0271074
26. Sharif N, Alzahrani KJ, Ahmed SN, Opu RR, Ahmed N, Talukder A, Nunia R, Chowdhury MS, Nodi IJ, Saha T, Zhang M (2021) Protective measures are associated with the reduction of transmission of COVID-19 in Bangladesh: a nationwide cross-sectional study. PLoS ONE 16(11):e0260287
27. Sharif N, Sharif N, Alzahrani KJ, Halawani IF, Alzahrani FM, Díez ID, Lipari V, Flores MA, Parvez AK, Dey SK (2023) Molecular epidemiology, transmission and clinical features of 2022-mpox outbreak: a systematic review. Health Sci Rep 6(10):e1603
28. Sharif N, Dey SK (2021) Impact of population density and weather on COVID-19 pandemic and SARS-CoV-2 mutation frequency in Bangladesh. Epidemiol Infect 149:e16
29. Sharif N, Sarkar MK, Ahmed SN, Ferdous RN, Nobel NU, Parvez AK, Talukder AA, Dey SK (2021) Environmental correlation and epidemiologic analysis of COVID-19 pandemic in ten regions in five continents. Heliyon e06576:1–9
30. WHO Coronavirus Disease (COVID-19) Dashboard. https://covid19.who.int//. Accessed on 17 Feb 2024
31. Priya SS, Cuce E, Sudhakar K (2021) A perspective of COVID 19 impact on global economy, energy and environment. Int J Sustain Eng 14(6):1290–1305
32. Bagchi B, Chatterjee S, Ghosh R, Dandapat D (2020) Impact of COVID-19 on global economy. In: Coronavirus outbreak and the great lockdown. SpringerBriefs in Economics. Springer, Singapore. https://doi.org/10.1007/978-981-15-7782-6_3
33. Shrestha N, Shad MY, Ulvi O, Khan MH, Karamehic-Muratovic A, Nguyen US, Baghbanzadeh M, Wardrup R, Aghamohammadi N, Cervantes D, Nahiduzzaman KM (2020) The impact of COVID-19 on globalization. One Health 11:100180
34. Kumar V, Alshazly H, Idris SA, Bourouis S (2021) Evaluating the impact of COVID-19 on society, environment, economy, and education. Sustainability 13(24):13642
35. Yeyati EL, Filippini F (2021) Social and economic impact of COVID-19. Brookings Institution. Accessed on 17 Feb 2024
36. Bambra C, Riordan R, Ford J, Matthews F (2020) The COVID-19 pandemic and health inequalities. J Epidemiol Community Health 74:964–968
37. Sutherland K, Chessman J, Zhao J, Sara G, Shetty A, Smith S, Went A, Dyson S, Levesque JF (2020) Impact of COVID-19 on healthcare activity in NSW Australia. Public Health Res Pract 30(4):e3042030
38. Chatterjee K, Chatterjee K, Kumar A, Shankar S (2020) Healthcare impact of COVID-19 epidemic in India: a stochastic mathematical model. Med J Arm Forces India 76(2):147–155
39. Khetrapal S, Bhatia R (2020) Impact of COVID-19 pandemic on health system & sustainable development goal 3. Indian J Med Res 151(5):395
40. Andraska EA, Alabi O, Dorsey C, Erben Y, Velazquez G, Franco-Mesa C, Sachdev U (2021) Health care disparities during the COVID-19 pandemic. In: Seminars in vascular surgery 34:82–88). WB Saunders
41. Okonkwo NE, Aguwa UT, Jang M, Barré IA, Page KR, Sullivan PS, Beyrer C, Baral S (2020) COVID-19 and the US response: accelerating health inequities. BMJ Evid Based Med. https://doi.org/10.1136/bmjebm-2020-111426

42. Watson MF, Bacigalupe G, Daneshpour M, Han WJ, Parra-Cardona R (2020) COVID-19 interconnectedness: health inequity, the climate crisis, and collective trauma. Fam Process 59(3):832–846
43. Benfer EA, Vlahov D, Long MY, Walker-Wells E, Pottenger JL, Gonsalves G, Keene DE (2021) Eviction, health inequity, and the spread of COVID-19: housing policy as a primary pandemic mitigation strategy. J Urban Health 98:1–2
44. Blume AW (2022) Health inequity and COVID-19. In: Colonialism and the COVID-19 Pandemic. Springer, Cham, pp 103–123
45. McGrail K, Morgan J, Siddiqi A (2022) Looking back and moving forward: addressing health inequities after COVID-19. Lancet Reg Health-Am 9:100232
46. Vásquez-Vera H, León-Gómez BB, Borrell C, Jacques-Aviñó C, López MJ, Medina-Perucha L, Pasarin M, Sánchez-Ledesma E, Pérez K (2023) Inequities in the distribution of COVID-19: an adaptation of WHO's conceptual framework. Gac Sanit 36:488–492
47. Lopez L, Hart LH, Katz MH (2021) Racial and ethnic health disparities related to COVID-19. JAMA 325(8):719–720
48. Rodier G, Greenspan AL, Hughes JM, Heymann DL (2007) Global public health security. Emerg Infect Dis 13(10):1447
49. Heymann DL, Chen L, Takemi K, Fidler DP, Tappero JW, Thomas MJ, Kenyon TA, Frieden TR, Yach D, Nishtar S, Kalache A (2015) Global health security: the wider lessons from the west African Ebola virus disease epidemic. The Lancet 385(9980):1884–1901

Nadim Sharif is a lecturer at the Department of Microbiology, Jahangirnagar University. He has completed both MS and BSc in Microbiology with first-class first positions from the Department of Microbiology, Jahangirnagar University. He has been awarded the Government Merit Scholarship for Outstanding performance in BSc. He has been working in molecular epidemiology of DNA and RNA viruses, public health, infectious diseases, and emerging viral diseases. He has in total fourty research articles in international scientific journals and fifteen international seminar papers. Now he is working with RNA viruses, namely, rotaviruses, noroviruses, chikungunya viruses, dengue viruses, SARS-CoV-2, and DNA viruses, e.g., adenoviruses and human bocaviruses. He is continuously contributing to the field of virology and public health. He is an editorial board member in PLOS ONE journal and review editor in Frontiers in Microbiology journal.

Shuvra Kanti Dey Ph.D., is a Professor at the Department of Microbiology, Jahangirnagar University. Before joining Jahangirnagar University as an assistant professor, he has worked in icddr, b. He has completed his MPH and Ph.D. from Tokyo University, Japan. He has been awarded the JSPS fellowship and another postdoc fellowship at Maryland University, USA. He has been working with epidemiology, public health, virology, infectious diseases, and emerging diseases for a long time. He has been an author and co-author of more than 70 original research articles and more than 40 seminar papers. He has been a peer reviewer of many well-known journals. Now he is working with emerging infectious RNA viruses, namely, rotaviruses, noroviruses, chikungunya viruses, dengue viruses, SARS-CoV-2, and DNA viruses, e.g., adenoviruses and human bocaviruses, and other bacterial pathogens with antibiotics resistant properties. He is continuously contributing to the field of virology, epidemiology, infectious diseases, and public health.

Global Disparities in HIV/AIDS Treatment

3

Tapen Sinha

> *"Where are the drugs? The drugs are where the disease is not. And where is the disease? The disease is where the drugs are not." Dr Peter Mugyenyi, HIV/AIDS researcher and physician, and director of Uganda's Joint Clinical Research Centre, July 2000, International AIDS Conference, Durban.*

Abstract

Unlike most other diseases, HIV/AIDS affects a large number of people in developing countries. It has no vaccine to prevent it, although a preventative medication is available. It is treatable with lifetime medication. Most affected countries cannot produce any of these drugs. Many patients are stigmatized for their sexual practices. The global disparities for HIV/AIDS arise from these reasons. Africa is most affected by the disease. Yet, it cannot produce HIV/AIDS medication. Four decades after the identification of the underlying virus, there is no vaccine on the horizon. We explain a possible underlying economic reason for this paradox. The future for most recipients of medication for HIV/AIDS remains precarious for political reasons.

Keywords

HIV/AIDS • Inequity • Africa • Patents • Vaccine • Pandemic

T. Sinha (✉)
Instituto Tecnologico Autonomo de Mexico (ITAM), Ciudad de México, Mexico
e-mail: tapen@itam.mx; tapen@cmi.ac.in

Chennai Mathematical Institute, Chennai, India

University of Nottingham, Nottingham, UK

B. W. K. Son (ed.), *The Landscape of Global Health Inequity*, Integrated Science 22,
https://doi.org/10.1007/978-3-031-60502-4_3

1 Introduction

A leading HIV researcher [1] introduced his findings: "This study confirms that the antibodies (in monkeys) are, at the structural and genetic levels, similar to the human antibody that we need as the foundation for a protective HIV vaccine". It was forty years ago in 1984, the root cause of AIDS was identified as HIV. Yet, forty years later, we are still grappling to find a vaccine with an efficacy level beyond 50% [2]. Contrast that with Covid. After the genetic composition of COVID-19 was identified on 11 January 2020, it took less than three months to develop mRNA vaccine for it. It had an efficacy level of over 90%.

By one estimate, the US Federal Government alone spent over $30 billion on COVID-19 vaccine development within three years [3]. The money spent on HIV/AIDS vaccines has been a small fraction of that. In the four decades since its identification (the virus has existed for a long time), at least 40 million people have died of HIV/AIDS.

Why do we see such a stark difference in how Covid-19 was dealt with versus HIV/AIDS? The answer lies in the kinds of people who are mostly affected by HIV/AIDS. COVID-19 affected developed and developing countries with equal force. The fatality is higher for older individuals. In developed countries, there is a higher proportion of older individuals. Thus, it hit the developed world proportionally more than the developing countries.

2 Historical Background

HIV/AIDS ranks right after the Spanish Flu in terms of global deaths in the past century. Covid-19 might eventually lead to a global death toll of ten million, but it is unlikely to reach forty million from HIV/AIDS. AIDS was recognized in 1981. But, identifying the HIV that causes AIDS took another three years to be recognized. It is now thought that HIV came out of a simian immunodeficiency virus (SIV) that affects monkeys in West Africa. The exact mechanism of the jump from monkeys to humans is still being debated, it was probably the railroad building that brought humans in close contact with SIV-infected chimpanzees. The CDC currently espouses that theory. About the origin, it noted, "HIV infection in humans came from a type of chimpanzee in Central Africa. Studies show that HIV may have jumped from chimpanzees to humans as far back as the late 1800s. The chimpanzee version of the virus is called the simian immunodeficiency virus. It was probably passed to humans when humans hunted these chimpanzees for meat and came in contact with their infected blood. Over decades, HIV slowly spread across Africa and later into other parts of the world. The virus has existed in the United States since at least the mid to late 1970s" [4].

The propagation of HIV is quite different from most other viruses. HIV cannot autonomously make copies of itself. To replicate HIV destroys the CD4 T lymphocyte (white blood) cells of the human immune system. Thus, HIV does not directly kill humans. However, the destruction of CD4 cells leads to a weakening of the immune system making the body vulnerable to other opportunistic infections. Deaths occur from pneumonia, cancer, and other diseases that cannot be stopped by the guardian CD4 cells. The very first reported cases of AIDS were all of Pneumocystis Pneumonia [5]. The CDC notes, "HIV uses the machinery of the CD4 cells to multiply and spread throughout the body. This process, which is carried out in seven steps or stages, is called the HIV life cycle. HIV medicines protect the immune system by blocking HIV at different stages of the HIV life cycle" [6].

HIV belongs to a class of viruses called the lentivirus (slow-moving virus). It takes between five to ten years for the seven stages of HIV to become full-blown AIDS. Thus, HIV/AIDS is a slow-moving virus that takes a decade to become a pandemic. On the other hand, influenza or Covid-19 are fast-moving viruses that become global in months. This time differential has an impact on the speed at which health authorities at the national or global levels act. A fortiori, it has an impact on the funding to treat or develop a vaccine for the virus.

3 How the Global Pandemic of HIV/AIDS Unfolded

Before airplane travel became common, the propagation of HIV was mostly local. People living in villages in remote parts of Africa did not have contact with people from outside their villages. The situation was similar in Europe a few generations earlier. The average distance traveled by a villager in a lifetime was less than seven kilometers up to the nineteenth century. Therefore, virulent pandemics which killed people quickly did not have a chance to proliferate far and wide. This is the principal reason for HIV/AIDS to become a global pandemic only in the 1980s. Similarly, the Ebola virus has circulated in Central Africa for decades. However, it became a global threat as a result of carriers moving fast across the globe by airplanes before manifesting the signs of the disease or dying.

If we examine the global propagation of HIV/AIDS (Fig. 1), we see two striking features.

First, the global deaths went from 330,000 in 1990 to over 1,800,000 in 2004. Then it started to fall until it hit 800,000 in 2019. It is important to remember that these are the midpoints of the estimates of each country and then added up. HIV/AIDS is such a stigmatized disease, in poorer countries, often the real cause of death is not recorded. Thus, there is substantial uncertainty about these estimates. The uncertainty is not symmetric around the mean because they are more likely to be underestimated.

Second, the deaths in Africa account for between 80 and 75% of the total number of deaths. Thus, without a doubt, the epicenter of the HIV/AIDS epidemic is in Africa. In particular, it is concentrated in Sub-Saharan Africa.

Therefore, most of the deaths from HIV/AIDS have been in (mostly) poor countries. On the other hand, the development of patented drugs to treat them has been in the developed world. This has created a huge asymmetry resulting in inequality that permeates much of our discussion below. Unlike COVID-19, the infection is

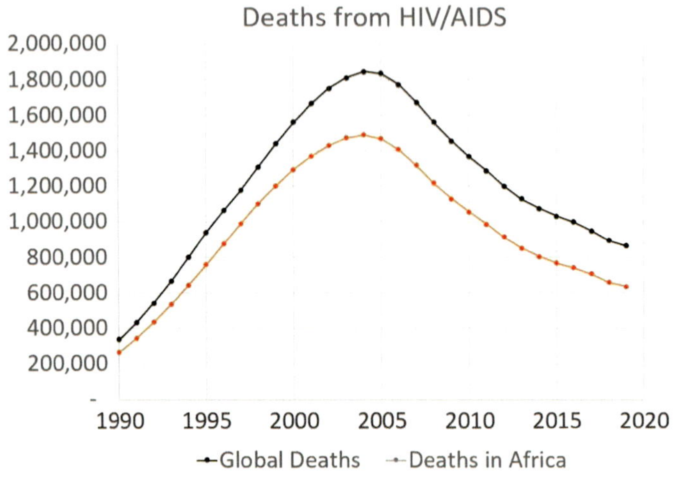

Fig. 1 Total number of deaths from HIV/AIDS between 1990 and 2019. *Source* World Bank Database on HIV/AIDS, 1990–2023

not distributed equally within a given country across all age groups. HIV/AIDS is (mostly) transmitted through (unprotected) sex. Hence, the affected population is (mostly) concentrated among with working age population.

In the US, in 2021, 70% of 32,000 new infections occurred among men who have sex with other men. Heterosexual contacts produced another 22% of new infections. An additional seven percent occurred among drug injectors [7]. In contrast, in South Africa, nearly 90% of all new infections were through heterosexual contacts [8].

4 The African Story of HIV/AIDS

In the previous section, we highlighted that the main drama of HIV/AIDS is playing out in Africa. If we examine specific countries with the highest burden of HIV/AIDS, the following picture emerges (Table 1).

The highest prevalence in the world in 2022 is in Eswatini (called Swaziland before)—a landlocked absolute monarchy in Southern Africa of 1.2 million people with a (nominal) per capita GDP of nearly four thousand US dollars. In the past two decades, Eswatini's economic growth has stalled. A large contributing factor has been the persistent high rate of prevalence of HIV/AIDS. Over 90% of all patients in Eswatini are aged between 15 and 49 [10].

A big contributor to this drain on growth is the scourge of HIV. The mechanism of the household-level economic impact is the following: The principal of the household is often the first victim. Then, other family members get infected. First comes the loss of income of the breadwinner, then the medical expenses. Subsequently, the women work less outside homes to take care of the sick person(s).

Table 1 Analysis of highest prevalence countries in the world

	Prevalence	GDPPC 2015	Cost 2015 (%)	Cumulative cost (%)
Eswatini	27.9	3600	3.90	39
Lesotho	20.9	1113	2.50	25
Botswana	18.6	6300	1.12	11
South Africa	18.3	6200	0.70	7
Namibia	11.8	4900	0.50	10
Zimbabwe	11.6	1400	2.10	42
Zambia	10.8	1300	2.50	25
Malawi	7.7	544	3.80	95
Equatorial Guinea	6.9	9700	0.30	3
Uganda	5.4	864	0.87	22

Source Own calculations based on the CIA Factbook of 2021 and updated calculations of [9]

Children do not get educated. Some get infected. The whole process produces a vicious cycle of poverty in the household. The same analysis applies to all the countries with high prevalence in Table 1.

Condon and Sinha [11] called it the "Zambian Dilemma": "Zambia has a high-prevalence, generalized HIV/AIDS epidemic. In 2006, Zambia's population was 11.5 million growing at the rate of 2.2% per year. The HIV prevalence rate among adults 15–49 years was 17%, an estimated 98,000 people died of AIDS, the number of children and adults (0–49 years) living with HIV was around 1.1 million and the number of AIDS orphans was estimated to be 710,000. Forty-six percent of the population was aged 0–15 years. The GDP of Zambia was USD 6.93 billion making the average per capita income around USD 600 in current dollars or USD 900 when adjusted for purchasing power parity. The income was not evenly distributed, so 64% of Zambians had expenditures below USD 1 per day. At 46% of the total population, there are over 5 million children in Zambia. Of them, there are 1 million child welfare needs for children affected by HIV/AIDS. Then, USD 1 billion would be required to meet the needs of these children. This would be equivalent to 14% of Zambia's GDP or 17% of total (public and private) consumption expenditure or 54% of total government expenditure, 48% of total imports of goods and services, and 309% of international grants provided to the government. And this is not a one-time expenditure."

The other dimension of the problem is illustrated in the two final columns of Table 1. The column entitled "Cost 2015" represents the cost to treat current HIV patients through anti-retroviral (ARV) drugs as a percentage of the GDP for 2015. The cumulative cost represents the total cost of treating those patients over their lifetimes as a percentage of the GDP applying a discount rate of 5%. These costs are high.

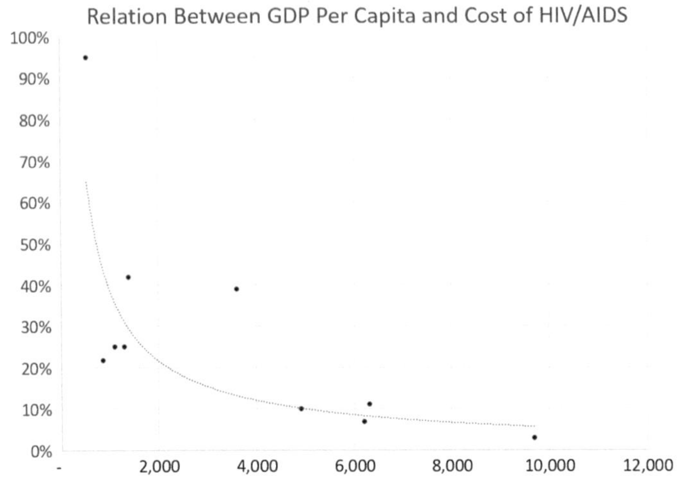

Fig. 2 Poverty and treatment cost of HIV/AIDS for the year 2021. *Source* Own calculations based on the CIA Factbook of 2021 and updated calculations of [9]

The poverty/HIV nexus becomes clearer from the following graph (Fig. 2). It shows that the cost of treating existing HIV patients (as a percentage of GDP) falls with the rise of the GDP per capita.

In summary, for most developed countries, new infections have dwindled through aggressive and effective treatment regimes. Deaths from HIV/AIDS are also on the way out. For example, Australia is on track to meet the UN goal of eliminating HIV/AIDS as a public health by 2030 [9]. For poorer countries in Africa, it is another matter. Propagation through heterosexual transmission is still killing over half a million people every year.

5 Global Political Economy of Pharmaceuticals

Most drugs to treat HIV/AIDS today are a product of research and development in the previous century. Theoretically, they should all be available at low cost in poorer countries under compulsory licensing. Under the World Trade Organization's (WTO) Agreement on Trade-Related Aspects of Intellectual Property Rights (TRIPS), a country can suspend patent rights and allow for compulsory licensing of patented drugs. Under compulsory licensing, the government can force a pharmaceutical company to produce a drug if it declares a national emergency for the use of the drug (TRIPS Article 31) [11].

In 2001, after the attack on the Twin Towers in New York City, the US Government suspected that anthrax was being weaponized by Iraq. There were 22 cases of anthrax (eleven inhalational and eleven cutaneous) were identified in the US. Five inhalational cases were fatal. The US and the Canadian governments used the threat of a potential bioterrorist attack to force Bayer—the largest maker of anthrax treatment drug ciprofloxacin—to lower the price to stockpile the drug. In reality, there was no actual evidence of anthrax being exported by a foreign country to the US or Canada. There was a perception of a threat by the governments. Yet, these governments were able to get Bayer to comply. Bayer agreed not to sue the Canadian government for violating its patent laws and to supply one million doses of ciprofloxacin within two days. Bayer agreed to "discount" the price both to the US and Canadian governments. It reduced the price by 40% [11].

Around the same time, in Doha, the TRIPS agreement was being negotiated among the members of the WTO. The final agreement contained the following provisions:

"We agree that the TRIPS Agreement does not and should not prevent members from taking measures to protect public health. Accordingly, while reiterating our commitment to the TRIPS Agreement, we affirm that the Agreement can and should be interpreted and implemented in a manner supportive of WTO members' right to protect public health and, in particular, to promote access to medicines for all. In this connection, we reaffirm the right of WTO members to use, to the full, the provisions in the TRIPS Agreement, which provide flexibility for this purpose" [12]. It was recognized that the wording was vague. The result was the TRIPS Declaration to clarify the application of TRIPS provisions to public health crises (including HIV/AIDS, malaria, and tuberculosis).

The WTO recognized that in the least developed countries, manufacturing the drugs domestically was never an option. The only option these countries had was to import generic versions of these drugs from other developing with large capacities (such as Brazil and India) to supply them. It was uncertain under the TRIPS Agreement whether such an export was legal. That discussion resulted in the Decision of the Council for TRIPS of 27 June 2002 (the Transition Period Decision) and a Waiver delaying the application of TRIPS patent obligations for pharmaceuticals in the least-developed countries until 2016. This became known as the Paragraph 6 provision.

> We recognize that WTO members with insufficient or no manufacturing capacities in the pharmaceutical sector could face difficulties in effectively using compulsory licensing under the TRIPS Agreement. We instruct the Council for TRIPS to find an expeditious solution to this problem and to report to the General Council before the end of 2002 [12]

6 The Political Economy of HIV/AIDS in a Developing Country

Suppose a small Latin American country buys a patented drug for a thousand patients for 10,000 dollars per year per person. The profit is nearly 10 million dollars if the marginal cost is extremely low. Suppose it spends a million dollars on the political parties (either directly or by subsidizing the education of the children of the politicians in the US), it can report 9 million dollars in profit to the US government. The US government can then tax those profits, generating revenue that can be channeled into foreign aid programs in that country's government to pressure the government to restrict the use of generic drugs as substitutes for the disease.

It seems like a win all around. The US Government gets extra tax money, the ruling elite of that Latin American country gets a cash infusion and a thousand patients benefit in that country. Where is the downside? The answer lies in what could have been done if a generic version of the drug was bought from a third country at a cost of a thousand dollars per person instead of ten thousand dollars per person. Thus, for the same amount of money (ten million), the country could have treated ten thousand patients instead of a thousand—a tenfold rise in the number of patients being treated. This example is not an idle one. This drama played out in real life during 2007–2008. For legal reasons, we cannot name the country in question.

Figure 3 illustrates how the pricing strategy played out for the ART treatment for HIV/AIDS. In early 2000, the drug cocktail was for over 10,439 dollars. Brazil's government laboratories produced and sold to the government-sponsored programs in Brazil the same cocktail for 2767 dollars in June 2000. Indian company Cipla started selling the cocktail for USD 800 in September 2000. By October 2000, the patented version started selling for 931 dollars. As generic companies kept on reducing the price, the patented companies followed. It is interesting to note that the price of the generic always hovered between 23 and 29% of the patented ones.

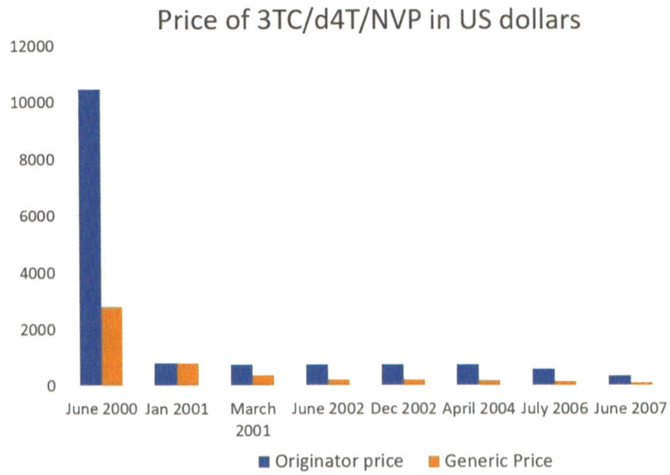

Fig. 3 Prices of HIV/AIDS drug 3TC/d4T/VP (combination drug). *Source* www.MSF.org. It was provided by James Love in 2005. He was instrumental in negotiating reduced drug prices and bringing generic drugs to countries where it was no longer necessary to sell patented drugs by the rules of the WTO

7 HIV: Vaccine Versus Treatment

Why is there no vaccine for HIV? The scientific answer is the following: Unlike other viruses, HIV integrates itself into the host DNA. Therefore, it limits the platform that can be used to develop a vaccine [13]. There is an economic side to the explanation. This was propounded by the Nobel Laureate in economics, Michael Kremer, and his coinvestigator Christopher M. Snyder. Their paper dates back to 2003 [14].

Their first paper was a theoretical one that contrasted the willingness to pay versus what can be charged in the market for a vaccine for a disease and a drug for the disease. They argue that a firm can generate more profits by drug development than by vaccine development. Below we summarize their argument and illustrate with one example they used.

A vaccine is always sold before consumers get the disease. Thus, the consumers differ considerably in disease risk. As a result, the firm cannot extract the consumer surplus (measured by the difference between what the consumer is willing to pay and what the firm can charge) with a uniform price. A drug, in contrast, is sold after consumers have contracted the disease. Thus, the firm can extract more consumer surplus from these relatively homogeneous consumers since the firm knows the buyer of the drug is infected (therefore willing to pay a higher than market price to save his/her life).

Result: The drug is more lucrative and hence the firm is biased toward developing the drug over the vaccine even if the vaccine is substantially more effective or lower cost. This result was illustrated with a simple example by Kremer and Snyder in [15]. "Suppose there are 100 total consumers, ninety of whom have a ten percent chance of contracting the disease and ten of whom have a 100% chance. Suppose consumers are risk-neutral and are willing to pay 100,000 to be cured of the disease if they contract it. A monopolist selling a vaccine could either charge 100,000 and sell to the ten high-risk consumers or charge 10,000 and sell to all 100 of them. Either way, the monopolist's revenue is 1,000,000. A monopolist selling a treatment would, in expectation, sell to the nineteen consumers contracting the disease (all ten of the high-risk consumers as well as an average of nine consumers from the low-risk group) for 100,000 for a total revenue of 1,900,000, almost twice the revenue from a vaccine" [15].

Kremer and Snyder [14] showed that for the same disease, a drug may be more lucrative than a vaccine. If the disease is contagious, the profit from the drug is even higher. In a follow-up paper, Kremer and Snyder [16] calibrated the situation with HIV data to conclude: "We find that a hypothetical drug monopolist would price an HIV drug so high that only 4% of the infected population worldwide would purchase, matching actual drug prices and quantities in the early 2000s before subsidies in low-income countries ramped up" [16].

8 National and International Initiatives Against HIV/AIDS

In 2022, the funds that went to fight against HIV/AIDS in low to middle-income countries amounted to 21 billion dollars (indexed to 2019). Of that, 12.5 billion came from public and private funds in those countries themselves. Another 4.8 billion came from the US in the form of bilateral agreements. Another 2.4 billion was from the Global Fund and 1.2 billion from other sources [17]. However, it is necessary to have 29.3 billion dollars per year to meet the needs (The path that ends AIDS: UNAIDS Global AIDS Update 2023. Geneva: Joint United Nations Program on HIV/AIDS). Yet, the available funds are shrinking over time (Fig. 4) [17].

The funds available rose from 16.6 billion dollars in 2010 to a peak of 23 billion in 2017. The composition of the source of funds has slowly changed over that time horizon. In 2010, slightly over half the funds came from domestic sources. The US funds stayed stable at around 22–23% and the Global Fund provided another 9–11%. Other funds have contributed 18% of the total in 2010. But the proportion has shrunk to 6% by 2022. The difference has been picked up by domestic sources. In other words, in 2022, 60% of funds came from domestic sources. However, the total has fallen from 23 billion to less than 21 billion. Therefore, the goal of bringing HIV/AIDS out of a public health threat by 2030 will not be realized. There is no possibility of getting a vaccine out on a global scale by 2030.

What exactly have these funds achieved? They have kept alive an estimated 20.8 million people who would have died in the absence of the drugs they get to keep them alive. This result can be seen in Fig. 5. The rollout of drugs started globally

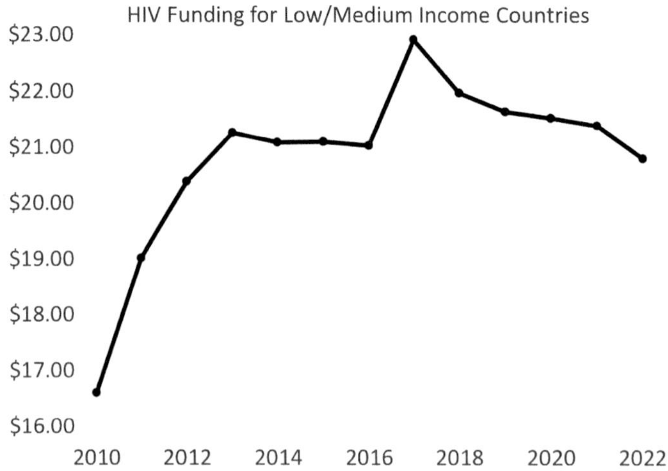

Fig. 4 Funding HIV/AIDS in low/medium income countries (in billions) of 2019 dollars. *Source* Kremer and Snyder [17]

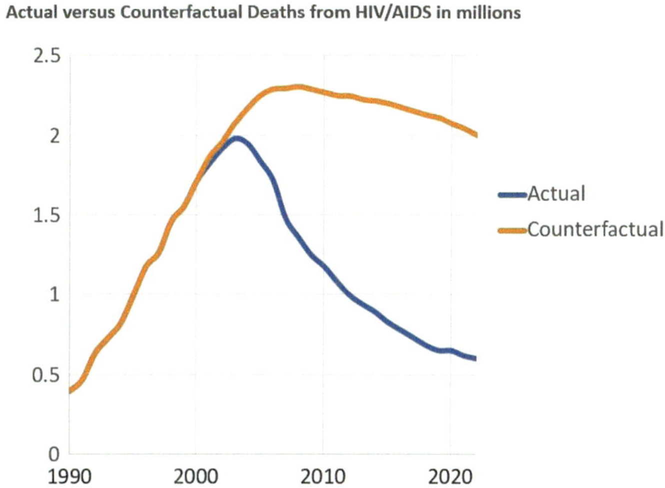

Fig. 5 Deaths avoided through the use of drugs for HIV/AIDS patients. *Source* UNAIDS 1990–2023

at scale after the United Nations set up the UNAIDS initiative. Smaller initiatives like the ones by Doctors without Frontiers started earlier. Around the same time, the US initiative called the PEPFAR (President's Emergency Plan for AIDS Relief) targeted HIV/AIDS. In addition, the Global Fund allocated money for HIV/AIDS. All of these projects were directed to (1) treatment for the patients, (2) prevention (for example, mother-to-child transmission), (3) education for stopping further propagation (such as making needles available for drug users) among other things. Many of these programs are targeted at people who have been discriminated against by societies across the world (such as sex workers, homosexual men, and drug users). Thus, allocating funds for such programs, especially in low-income countries, has been fraught with controversies.

Since 20 odd billion dollars (indexed to 2019) is keeping 20 odd million people alive, we can reframe the initiatives to be spending a thousand dollars (indexed to 2019) per person per year. Table 1 illustrates that for half of the countries with the highest prevalence, nearly half of the population in each country earns less than that in an entire year. Therefore, it is not easy for the respective countries to justify such an ongoing expenditure. Some have suggested those resources might be better spent on strategies to reduce transmission. Oster (2007) found that the transmission rate in Africa is three times that of the US. She found that chlamydia, gonorrhea, and syphilis infections are highly correlated with the presence of HIV/AIDS. Her prescription was to divert money to treat these other three diseases to tackle the future spread of HIV/AIDS. The PEPFAR has been heavily touted as the lifeline for 20 million people [18]. However, as we saw above, it accounts for a quarter of the total allocation for HIV/AIDS. There is no doubt that it has helped millions. But the drugs used were the patented versions. That meant the cost was actually three times more than the generic versions. To put it differently, had the money been used to buy generics, it could have helped three times more people than it did. Why did PEPFAR not use the generics? The terms were dictated by the US pharmaceutical companies. They are among the largest donors in the US political system. This loops back to Sect. 6 with the example of that Latin American country.

9 Rich Countries Have Other Levers

Figure 6 illustrates two facts: The global prevalence of HIV and the proportion of HIV patients with access to the medication for treating it. Globally, the rise of HIV as a proportion of the total number of people seems to be peaking at around 0.5%. However, the absolute number of patients is still rising because, in the past two decades, the world population has grown from nearly 6 billion to over 8 billion. Figure 6 also shows that the proportion of HIV patients with access to medicine has grown from under 10% in 2004 to 78% in 2022.

We are in a situation where 40% of the funding comes from rich countries. In particular, the US PEPFAR program provides at least half of that. PEPFAR requires a Congressional authorization every year from the US. Until now, there has been bipartisan support for it. However, the majority in Congress has expressed their unwillingness to support it any further in late 2023. Thus, a good chunk of funding is a politically driven process making future funding uncertain.

In addition, the US government has shown that it is willing to use trade as a weapon against countries that try to circumvent the price monopoly of HIV drugs.

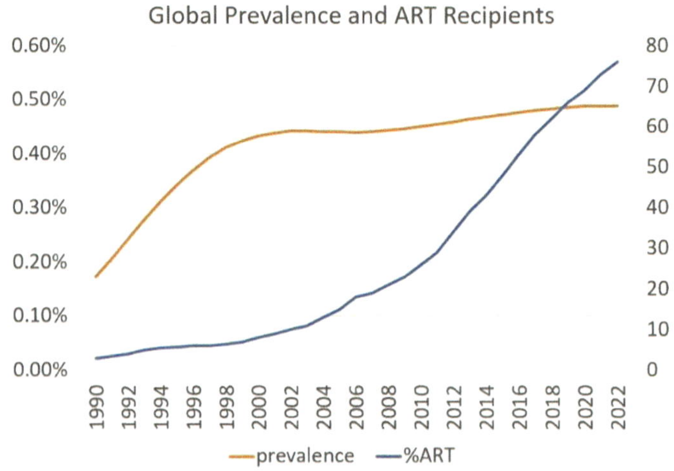

Fig. 6 Global prevalence (left scale) and percentage of patients receiving drugs (right scale). *Source* UNAIDS database at UNAIDS.org 1990–2022

In 2007, the U.S. Trade Representative's Office (USTR) signaled its willingness to fight a trade war with Thailand in a dispute about compulsory licensing for HIV drugs even though Thailand was compliant with the TRIPS provisions [19]. This example shows why poorer countries are often unwilling to use the legal means they can pursue. They may provoke consequences in other areas independent of the main dispute.

10 Conclusions

Dr. Peter Mugyenyi (whom we quote at the beginning) put his finger right on the pulse. For HIV/AIDS, the biggest burden of the disease is on poor people mainly in Africa. The mismatch between where the drugs are and where the disease is still staring at us a quarter of a century after he uttered those words. Vaccines for other diseases are not manufactured in Africa either. Of all the vaccines provided in Africa, less than one percent is produced in Africa [20].

If the governments of rich countries are serious about a vaccine against HIV/AIDS, they cannot rely on private pharmaceuticals to provide it. The financial incentive to develop a vaccine is absent in the private sector. There is a need for the governments of richer countries to fund such vaccines. Given that diseases like HIV/AIDS are contagious, the incentive to develop such vaccines is not merely charitable. Such diseases will arrive at the doorstep of all countries in the future as Covid-19 vividly illustrates.

References

1. Saunders K, Counts J, Thakur V, Monefiori AP, Haynes B (2024) Vaccine induction of CD4-mimicking HIV-1 broadly neutralizing antibody precursors in macaques. Cell 187:79–94. https://doi.org/10.1016/j.cell.2023.12.002
2. DukeHealth (2024) HIV vaccine takes step forward with confirmation of neutralizing antibodies. https://corporate.dukehealth.org/news/hiv-vaccine-takes-step-forward-confirmation-neutralizing-antibodies. Accessed on 3 Jan 2024
3. Kates J, Cox C, Michaud J (2023) how much could COVID-19 vaccines cost the U.S. after commercialization? https://www.kff.org/coronavirus-covid-19/issue-brief/how-much-could-covid-19-vaccines-cost-the-u-s-after-commercialization/. Accessed on 24 Dec 2023
4. CDC (2022) Where did HIV come from? https://www.cdc.gov/hiv/basics/whatishiv.html. Accessed on 13 Nov 2023
5. CDC (1981) Pneumocystis Pneumonia—Los Angeles. MMWR Weekly June 5, 1981 30(21):1–3 https://www.cdc.gov/mmwr/preview/mmwrhtml/june_5.htm Accessed on 4 Jan 2024
6. HIV Info (2021) The HIV Life Cycle. https://hivinfo.nih.gov/understanding-hiv/fact-sheets/hiv-life-cycle. Accessed on 6 Jan 2024
7. HIV.Gov. (Dec. 7, 2023) U.S. Statistics. https://www.hiv.gov/hiv-basics/overview/data-and-trends/statistics. Accessed on 28 Feb 2024
8. Malahleha M, Laher F, Janes DA (2023) Risk factors associated with HIV acquisition in males participating in HIV vaccine efficacy trials in South Africa. AIDS Behav 27:3027–3037. https://doi.org/10.1007/s10461-023-04025-z

9. Collier P, Sterck O (2018) The moral and fiscal implications of antiretroviral therapies for HIV in Africa. Oxf Econ Pap 70(2):353–374. https://doi.org/10.1093/oep/gpy001
10. Bollinger L, Stover J (1999) The economic impact of AIDS in Swaziland. https://pdf.usaid.gov/pdf_docs/pnacm922.pdf. Accessed on 19 Dec 2023
11. Condon B, Sinha T (2008) Global lessons from the AIDS pandemic economic, financial, legal and political implications, Springer.
12. WTO (2003) The Doha declarations. https://www.wto.org/english/res_e/booksp_e/ddec_e.pdf. Accessed on 2 Dec 2023
13. McGregor S, O'Donnel D (2023) Australia is close to eliminating HIV transmission, but not there yet, the diplomat, https://thediplomat.com/2023/12/australia-is-close-to-eliminating-hiv-transmission-but-not-there-yet/. Accessed on 11 Dec 2023
14. Kremer M, Snyder CM (2003) Why are drugs more profitable than vaccines? NBER Working Paper No. 9833.
15. Doherty Institute (2023) Why a cure for HIV remains elusive. https://www.doherty.edu.au/news-events/news/why-a-cure-for-hiv-remains-elusive. Accessed on 2 Jan 2024
16. Kremer M, Snyder CM (2015) Preventives versus treatments get access. Q J Econ 130(3):1167–1239. https://doi.org/10.1093/qje/qjv012
17. Kremer M, Snyder CM (2018) Preventives versus treatments redux: tighter bounds on distortions in innovation incentives with an application to the global demand for HIV pharmaceuticals. Rev Industr Organiz Industr Organiz Soc 53(1):235–273
18. UNAIDS (2024) Financial dashboard. https://hivfinancial.unaids.org/hivfinancialdashboards.html. Accessed on 3 Jan 2024
19. MSF (2007) Thailand turns giant pharma killer. https://www.msf.org/thailand-turns-giant-pharma-killer. Accessed on 9 Jan 2024
20. Nweneka CV, Disu T (2022) The future of vaccine manufacturing in Africa. Foresight Africa. https://www.brookings.edu/articles/the-future-of-vaccine-manufacturing-in-africa/
21. KFF Global Health Policy (2023) https://www.kff.org/global-health-policy/fact-sheet/the-u-s-presidents-emergency-plan-for-aids-relief-pepfar/. Accessed on 3 Jan 2023

The Inequity of Cancer Treatment Services in Kenya

4

Francis Wambalaba and Akosa Wambalaba

Abstract

In Kenya, cancer stands as the third leading cause of death, following infectious diseases and cardiovascular conditions. The period from 2012 to 2018 witnessed an alarming surge of nearly 16% in annual cancer mortality (Pankaj et al. in Cancer on the Global Stage: Incidence and Cancer-Related Mortality in Kenya. WHO Region, Africa The ASCO Post 2021 [1]). This increase is attributed to factors such as physical inactivity, unhealthy dietary patterns, overweight/obesity, HIV/AIDS, detrimental alcohol consumption, substance abuse, and environmental shifts. Prevalent cancer types in Kenya encompass cervical, breast, esophageal, and prostate cancers, along with others like head and neck, colon and rectum, stomach, liver, and soft tissue sarcomas. The grim reality unfolds particularly for over 70% of urban Kenyans residing in informal settlements, where health outcomes are dire. Morbidity and mortality rates in these areas surpass those in other parts of the country, aggravated by inadequate sector and service management, extensive inefficiencies, and corruption. The cancer scenario in rural areas compounds health challenges, marked by disparities such as sparse health facility distribution and financial constraints hindering access to private healthcare (Kabia et al. in Int J Equity Health 18:98,

This study was funded by the Rockefeller Foundation through the Africa Cancer Foundation (ACF) with contributions from Prof. Anyang Nyongo, Ms Dorothy Nyogo, Prof. Damary Sikalieh, and Ms Joy Wasike. A medical student, Mr. Otioli Wambalaba helped with registry card data collection. It was reviewed and edited by Dr. Barbara W. K. Son.

F. Wambalaba (✉)
United States International University, Nairobi, Kenya
e-mail: fwambalaba@usiu.ac.ke

A. Wambalaba
School of Humanities and Social Sciences, United States International University, Nairobi, Kenya
e-mail: awambalaba@usiu.ac.ke

B. W. K. Son (ed.), *The Landscape of Global Health Inequity*, Integrated Science 22,
https://doi.org/10.1007/978-3-031-60502-4_4

2019 [2]; Makau-Barasa et al. in Health Res Policy Sys 18:2, 2020 [3]). Ultimately, several demand side and supply side barriers exacerbate inequitable Access, Proximity, and Availability (APA) to cancer diagnosis and treatment. This chapter delves into the disparities associated with cancer and provides policy recommendations for addressing them.

Keywords

Cancer prevalence • Cancer screening • Cancer diagnosis and treatment • Cancer patients experience

1 Introduction

Non-communicable diseases cause more deaths in low- and middle-income countries (LMIC) than infectious diseases combined. In Africa, knowledge about cancers is unfortunately limited, and the disease has not received sufficient attention. Kenya lacks comprehensive national data to accurately depict the cancer situation. Cancer ranks as the third leading cause of death in the country, following infectious diseases and cardiovascular disease [4]. Urban areas in Kenya, where over 70% of the population resides in informal settlements, face alarming health outcomes. Inequitable health challenges concerning cancer are also prevalent in rural populations, characterized by sparse health facilities and financial constraints hindering access to private healthcare [2, 3].

Cancer arises from a combination of internal and external risk factors, triggered by their interaction in a sequence. Exposure to cancer-causing agents can occur in the environment or through lifestyle choices. This chapter addresses inequities in cancer treatment from both demand and supply perspectives. It delves into demand-side barriers such as socioeconomic, cultural, transportation, geographic, and health literacy factors. On the supply side, constraints related to cancer diagnosis, treatment, and resources are examined. To address impediments to quality healthcare access for marginalized segments of society, a robust cancer intervention strategy is crucial. By scrutinizing both demand-side and supply-side barriers, we offer insights into the interconnected elements of cancer treatment. This paves the way for targeted interventions that address and alleviate inequities in cancer treatment.

2 Cancer Treatment Inequity: Demand Side Issues

2.1 Socioeconomic Barriers

The fear of financial constraints and the potential for abnormal results indeed pose significant obstacles to achieving equity in cancer treatment. These challenges contribute to disparities in access to care and outcomes among different socio-economic groups. With the national health insurance scheme enrolling less than 20% of the population, accessing healthcare remains a challenge due to socioeconomic barriers, as many Kenyans only enroll when already afflicted [5]. Affordability issues discourage individuals from seeking medical care and knowing their health status. The majority of Kenyans face difficulties accessing childhood cancer care and adequate cancer treatment services due to poor socioeconomic status. Financial challenges, particularly among the economically disadvantaged, often result in delayed treatment for childhood cancer patients, leading to unfavorable outcomes. The protracted and costly nature of cancer treatment, especially for advanced cases, contributes to the financial burden. Delayed and futile referrals escalate the need for intensive and expensive hospital treatment, disrupting livelihoods significantly. A 2018 household survey revealed that 8% of Kenyan households experienced catastrophic health expenditures, pushing over 1 million people into poverty annually [6].

Many patients cannot complete their prescribed treatment due to personal financial bankruptcy, prompting some to advocate prioritizing funds for cancers with better outcomes, such as breast and colon cancer, over advanced cancers with poorer prognoses like esophageal and pancreatic cancer [7]. The inadequacies of the national health insurance plan further exacerbate the problem, preventing patients from affording essential medical services [8]. The state-run National Hospital Insurance Fund contributes only 25,000 Kenyan shillings (equivalent to US $250) per patient for cancer care, a significantly lower amount than the actual cost of treatments. The substantial costs linked to cancer treatment, reach millions of Kenyan shillings. Consequently, patients often resort to community fundraising from friends, family, and well-wishers or may even forego treatment altogether [9]. Collaborative efforts between government and non-governmental organizations are crucial in developing effective drug dispensing, distribution, and payment mechanisms, providing support for economically disadvantaged patients.

2.2 Cultural Barriers

Cultural barriers can indeed play a significant role in impeding cancer treatment equity in Kenya, as they affect various aspects of healthcare-seeking behavior, awareness, and treatment adherence, with prevailing myths such as the belief that "cancer is caused by curses from ancestors" [5]. These barriers pose substantial challenges, particularly in the context of early detection. Cultural misconceptions contribute to a reluctance to seek timely medical interventions. There exists a societal assumption that children do not experience cancer, possibly due to the historical association of cancer with factors like smoking, unhealthy lifestyles, and limited reported cases in children compared to adults. The incidence of childhood cancer in Africa is harder to study compared to that in adults, further complicating efforts in cancer control and care.

As per the Kenya National Cancer Screening Guidelines [10], cervical cancer ranks as the second leading cause of death in women of reproductive age, with breast cancer taking the lead. Breast cancer is identified as the primary cause of death for women aged 35 to 55 in Kenya, affecting one in nine women, and survival rates are relatively low. The apprehensions associated with breast cancer encompass concerns about surgery, mortality, body image alterations, and impacts on sexuality. Addressing these fears requires disseminating information and knowledge to empower women to make informed decisions about their care. To overcome these cultural barriers, there is a crucial need for community sensitization and education initiatives aimed at promoting early detection of breast lumps. By providing accurate information and fostering awareness, it is possible to dispel myths and encourage proactive healthcare-seeking behaviors within the Kenyan community, thereby improving outcomes for individuals facing cancer-related challenges.

2.3 Transportation and Geographic Barriers

Transportation and geographical barriers pose significant challenges to cancer treatment equity in Kenya. The concentration of most cancer care services within a 5 km radius in Nairobi creates inaccessibility for economically disadvantaged patients residing outside the city. This situation becomes particularly problematic for those who struggle to afford both the cost of treatment and the associated travel and accommodation expenses. Given that approximately 78% of the Kenyan population resides in rural areas, many individuals in need of cancer care find themselves having to travel up to 600 km to access services [5]. Many patients from Western Kenya and the Rift Valley have to travel across the border to Uganda's Mulago Hospital for treatment [5]. Except for Kenyatta National Hospital, most cancer care facilities are private, and, apart from Pandya Hospital, the majority are located in Nairobi. This geographic concentration has implications for both the affordability of services for the economically disadvantaged and the accessibility of care in rural areas and distant towns from Nairobi [3]. The need for cancer treatment forces patients to navigate significant distances, incurring additional financial burdens and logistical challenges. Addressing these transportation and geographical barriers is essential to ensure equitable access to cancer care for all Kenyan residents, irrespective of their location and financial circumstances. Efforts should be directed towards decentralizing cancer care services and improving transportation infrastructure or subsidizing it to enhance the reach and availability of treatment options for a broader population.

2.4 Limited Health Literacy

Health literacy refers to an individual's capacity to access, comprehend, and process fundamental health information and services, enabling them to make informed decisions about their well-being. In Kenya, limited health literacy presents challenges to cancer treatment equity and prevention initiatives. A pervasive lack of awareness and accurate information regarding cancer hampers the progress of cancer treatment in the country, leading to delayed diagnoses, with many patients presenting at hospitals during advanced stages of the disease [11]. The primary obstacle to cervical cancer screening is the insufficient awareness of cervical cancer as a significant health threat among women [12]. The constrained health literacy obstructs patients' decision-making processes, impacting the effectiveness of treatment and follow-up care.

To address these challenges, enhancing community-based actions, particularly by reinforcing established women's or youth groups, and in particular engaging men in similar groups where experiences and information can be shared, would play a pivotal role in controlling especially the top three—breast and cervical cancer among women and prostate cancer among men in Kenya. Empowering community healthcare workers, local leaders, religious institutions, and other community organizations as primary channels of communication is essential. Strengthening their awareness of the urgency of cancer management and prevention would be invaluable. Furthermore, considering the widespread adoption of mobile phone technology among Kenyans, leveraging short text messages (SMS) alerts and apps can effectively disseminate pertinent health information, thereby bolstering health literacy [13].

3 Cancer Treatment Inequity: Supply Side Issues

3.1 Cancer Screening and Treatment Constraints

In Kenya, the primary challenges associated with cancer screening and treatment revolve around three key issues, Accessibility, Proximity, and Availability as described in the Cancer APA Barriers Framework below [14] (Table 1).

Table 1 The cancer APA barriers framework

#	Barrier	Description
1	Accessibility	The predominant provision of screening and treatment services by private hospitals has made these services more expensive as they lack subsidies. This financial burden limits accessibility for a significant portion of the population
2	Proximity	The concentration of most screening and treatment services in Nairobi poses a geographical challenge. The distance from these services to various regions in the country makes it difficult for many individuals to access timely and convenient care
3	Availability	The accessibility of screening and treatment options is contingent upon the presence of essential equipment and drugs. The limited availability of these resources further compounds the challenges faced by individuals seeking cancer care

Source Derived from Wambalaba et al. [5]

Accessing cancer screening and treatment is one of the major hurdles Kenyan cancer patients face. Severe disparities in the distribution of cancer treatment centers contribute to inequities. At the time of the study, only four cancer screening centers existed in Kenya, all located in Nairobi, and three of them were privately owned. Kenyatta National Hospital, the sole public center in the country treating cancer, bore the brunt of this burden, hosting the majority of Kenya's oncology expertise and technology [15]. The hospital's pediatric oncology unit, with a mere 28 beds, catered to only 25% of pediatric cancer cases, necessitating the accommodation of the rest in other wards. Kenyatta National Hospital was grappling with an overwhelming caseload, evident in a two-month waiting list for new patients. Given the exorbitant costs of private cancer care, many Kenyans resort to seeking treatment in other countries, notably India, due to the prohibitive expenses. For instance, the cost of one week of radiation treatment at Kenyatta National Hospital was 2000 shillings ($24), while private hospitals like Aga Khan University Hospital, MP Shah Hospital, and Nairobi Hospital charge a staggering 40,000 shillings ($475) for the same treatment [5].

The government initiated the implementation of cervical cancer screening at selected HIV care centers, employing a method known as visual inspection with acetic acid. This approach was deemed cost-effective, particularly for individuals with limited financial means, in contrast to the more widely used Pap smear test. The Pap smear test requires highly trained health providers, access to specialized equipment and supplies, a cytology laboratory, as well as arrangements for communicating results to screened women and confirming diagnoses in a medical facility [16]. Unfortunately, many women were diagnosed with cervical cancer at advanced stages, presenting challenges for curative treatment, often rendering it difficult or even impossible. The adoption of cervical cancer screening remained low, with a mere 16% uptake, and for breast cancer, the rate was even lower at 12% [17]. According to the Kenya Breast Health Program, a staggering 95% of women in Kenya have never undergone a clinical breast examination due to the limited availability of screening mammography in the country. Late diagnosis emerges as the primary contributor to high mortality rates, with estimates indicating that 70 to 90% of breast cancer patients seek consultations when already in stage 3 or stage 4 of the disease [18, 19].

Among the prominent types of childhood cancers, blood cancers like leukemia and lymphomas, including Burkitt's lymphoma, exhibit favorable responses to early detection and chemotherapy. However, the outlook is less optimistic in sub-Saharan Africa, where Burkitt's lymphoma is the most prevalent childhood cancer. Burkitt's lymphoma accounts for half of the childhood tumors treated at regional hospitals in Kenya and Uganda. The units dedicated to childhood cancer care confront challenges related to limited access for patients in remote areas and a scarcity of specialized treatment centers. The diagnostic process is intricate, and the expense of treatment poses a significant hurdle. To mitigate disparities in cancer treatment, particularly affecting economically disadvantaged individuals, there is a pressing need for additional public resources. These resources would be instrumental in facilitating subsidized services for patients, enabling appropriate screening, and supporting the provision of essential treatments [19].

3.2 Resource Constraints

Disparities in the availability and sophistication of medical equipment and technologies for cancer diagnosis and treatment contribute to inequities in the quality of cancer care. In Kenya, a significant number of cancer cases often go undiagnosed or are misdiagnosed, a challenge linked to insufficient healthcare resources. Diagnostic services are predominantly concentrated in Nairobi and major cities but are constrained in terms of capacity [3]. Access to cancer drugs is inconsistent across many public hospitals, with some essential pain management medications being notably scarce. Kenya faces a deficiency of modernized equipment to effectively address cancer, resulting in limited exposure for medical personnel to emerging technologies. For instance, the Cobalt 60 machines utilized at Kenyatta National Hospital belong to an older generation, featuring limited capabilities for treating complex cases optimally. While private hospitals like Aga Khan or Nairobi Hospital boast advanced machines, their accessibility is restricted due to their high costs. For a time, the Cancer Care Centre near MP Shah Hospital in Nairobi stood as the sole facility in the country equipped with newer technologies like Linear Acceleration. Although this technology offered increased accuracy and handled a greater workload compared to the Cobalt 60, it came with higher expenses, necessitated a more skilled workforce, and imposed substantial maintenance demands [5].

3.3 Medical Personnel Constraints

Limited specialized human resource capacity stands out as a significant factor contributing to the inequity in cancer treatment in Kenya. The scarcity of cancer specialists, predominantly concentrated in a handful of health facilities in Nairobi, creates a substantial barrier for the majority of the population seeking access to cancer treatment services. This results in prolonged waiting times, allowing some previously manageable tumors to progress to incurable stages. The escalating prevalence of cancer has placed a strain on doctors, stretching their capacities beyond sustainable limits. Many medical professionals, particularly those trained in blood diseases and surgery, find themselves engaged in cancer treatment due to the overwhelming number of cancer patients surpassing the limited personnel resources in Kenyan hospitals [5].

The shortage of qualified healthcare professionals, including oncologists and specialized staff, further restricts the ability to deliver timely and effective cancer care. Addressing this challenge requires an increase in the number of pathologists and oncologists in peripheral hospitals to facilitate early detection and prompt treatment of cancer [4]. Additionally, a commitment from government, public, and private teaching hospitals is essential to expanding the incorporation of cancer education into basic medical and nursing training programs. is designed to bolster both the quantity and quality of healthcare professionals, ensuring they are well-equipped to alleviate the inequities in cancer treatment in Kenya.

4 Conclusions and Recommendations

Addressing both demand and supply-side barriers that contribute to inequities in cancer treatment in Kenya provides insights into the interconnected elements of cancer care. This understanding lays the foundation for targeted policy interventions aimed at alleviating disparities in cancer treatment. The health system in Kenya faces challenges, characterized by poorly equipped hospitals, a low doctor-to-patient ratio, insufficient skills among service providers, limited treatment facilities following pre-cancer or cancer diagnoses, and a low rate of cancer screening. Reducing inequities in access to and improving the quality of cancer screening in Kenya faces additional hurdles. These challenges include limited access to affordable cancer treatments, low community awareness regarding the importance of screening, limited knowledge of common cancer symptoms, inappropriate utilization of human resources, limited research and data to support policy formulation, and the concentration of cancer services in primarily Nairobi [2]. Addressing these multifaceted challenges is crucial for fostering a more equitable and accessible cancer treatment in Kenya.

Recognizing the considerable demand in the face of limited capacity, policymakers and practitioners are urged to meticulously assess the challenges associated with cancer and explore effective mitigation strategies. In July 2020, the Ministry of Health took a significant step by launching the New Cancer Control Strategy, a national initiative aimed at addressing the escalating burden of the disease in the country [1]. Drawing on the successful performance of private hospitals utilizing outsourced resources, it is recommended that a similar model be promoted within public facilities to leverage economies of scale, particularly beneficial for smaller facilities. Additional recommended strategies encompass the continuous updating of Cancer Treatment Guidelines, providing support for cancer pilot studies, intensifying efforts to address resource disparities across regions, enhancing cancer foundations (such as the Africa Cancer Foundation), and expanding cancer treatment coverage through the National Hospital Insurance Fund. These recommendations collectively aim to mitigate cancer treatment inequity in Kenya.

References

1. Pankaj GJ, Hope C, Are C, Rooprai G (2021) Cancer on the global stage: incidence and cancer-related mortality in Kenya, 25 Feb, WHO Region, Africa, The ASCO Post
2. Kabia E, Mbau R, Oyando R et al (2019) We are called the et cetera: experiences of the poor with health financing reforms that target them in Kenya. Int J Equity Health 18:98. https://doi.org/10.1186/s12939-019-1006-2

3. Makau-Barasa LK, Greene S, Othieno-Abinya NA et al (2020) A review of Kenya's cancer policies to improve access to cancer testing and treatment in the country. Health Res Policy Sys 18:2. https://doi.org/10.1186/s12961-019-0506-2
4. Topazian H, Cira M, Dawsey SM, Kibachio J, Kocholla L, Wangai M, Welch J, Williams MJ, Duncan K, Galassi A (2016) Joining forces to overcome cancer: the Kenya cancer research and control stakeholder program. J Cancer Policy March 1(7):36–41. https://doi.org/10.1016/j.jcpo.2015.12.001
5. Wambalaba FW, Son B, Wambalaba AE, Nyong'o D, Nyong'o A (2019) Prevalence and capacity of cancer diagnostics and treatment: a demand and supply survey of health-care facilities in Kenya. Cancer Control26(1):1073274819886930. https://doi.org/10.1177/1073274819886930
6. Ilinca S, Di Giorgio L, Salari P et al (2019) Socio-economic inequality and inequity in use of health care services in Kenya: evidence from the fourth Kenya household health expenditure and utilization survey. Int J Equity Health 18:196. https://doi.org/10.1186/s12939-019-1106-z
7. WHO (2022) Addressing inequities in breast cancer treatment in sub-Saharan Africa: insights from a breast cancer surgeon in Nairobi. https://www.who.int/news-room/feature-stories/detail/addressing-inequities-in-breast-cancer-treatment-in-sub-saharan-africa--insights-from-a-breast-cancer-surgeon-in-nairobi. Accessed 3 Feb 2024
8. Chelangat M (2023) KMPDU faults new social health insurance Act, Nov 10, Nation News. https://nation.africa/kenya/news/kmpdu-faults-new-social-health-insurance-act--4429868 Accessed 1 Feb. 2024
9. Ramrayka L (2018) Improving cancer in Kenya: a pathway to hope, March 15. www.roche.com. Accessed 3 Feb 2021
10. Kenya Ministry of Health (2018) Kenya national cancer screening guidelines. November, Nairobi. www.health.go.ke. Accessed 1 Feb 2024
11. Makau-Barasa LK, Greene SB, Othieno-Abinya NA, Wheeler S, Skinner A, Bennett AV (2018) Improving access to cancer testing and treatment in Kenya. J Glob Oncol 4:1–8. https://doi.org/10.1200/JGO.2017.010124
12. Jemal A, Brawley OW (2019) Increasing cancer awareness and prevention in Africa. Ecancermedicalscience 13:939. https://doi.org/10.3332/ecancer.2019.939
13. Ngaruiya C, Oti S, Van de Vijver S, Kyobutungi C, Free C (2019) Target women: equity in access to mHealth technology in a non-communicable disease care intervention in Kenya. PLoS ONE 14(9):e0220834
14. Wambalaba F, Son B, Wambalaba A, Nyong'o D, and Nyong'o A (2019) Prevalence and capacity of cancer diagnostics and treatment: a demand and supply survey of health-care facilities in Kenya. Cancer Control. https://journals.sagepub.com/doi/full/ https://doi.org/10.1177/1073274819886930
15. Makau-Barasa LK, Greene SB, Othieno-Abinya NA, Wheeler S, Skinner A, Bennett AV (2017) Improving access to cancer testing and treatment in Kenya. JCO Global Oncol https://ascopubs.org/journal/go. Accessed 4 Feb 2024
16. Munoru F, Gitonga L, Muraya M (2019) Integration of cervical cancer screening services in the routine examinations offered in the Kenyan health facilities: a systematic review. Open J Obstet Gynecol 9(5):656–668
17. Kagiri H, Kiarie H, Karagu A (2020) Assessment of demand versus supply of cancer services on Kenya. Ministry of Health. Countdown 2030.Org
18. Kisiangani J, Baliddawa J, Marinda P et al. (2018) Determinants of breast cancer early detection for cues to expanded control and care: the lived experiences among women from Western Kenya. BMC Women's Health 18:81. https://doi.org/10.1186/s12905-018-0571-7
19. Antabe R, Kansanga M, Sano Y et al (2020) Utilization of breast cancer screening in Kenya: what are the determinants? BMC Health Serv Res 20:228

Prof. Francis Wambalaba is a development economist at United States International University, Nairobi, Kenya studying health as a development issue.

Ms. Akosa Wambalaba is in the School of Humanities and Social Sciences at United States International University, Nairobi, Kenya. She teaches Ethical Values and Integrity and has been researching knowledge transfer.

COVID-19 Lessons and Medical Inequities in Africa

5

Abdiqani Sheikh Omar

Abstract

The COVID-19 pandemic, a global crisis of unprecedented proportions, has illuminated and exacerbated the longstanding issue of medical inequalities in Africa. This chapter delves into the multifaceted dimensions of this challenge, offering a comprehensive examination of the pandemic's impact on the continent, the underlying factors contributing to healthcare disparities, and the lessons drawn from these experiences. We begin by contextualizing the healthcare landscape in Africa and exploring historical factors that have shaped its development. We then dissect the African response to COVID-19, highlighting both successes and challenges faced by nations with varying levels of resources and infrastructure. Our analysis extends to the economic repercussions of the pandemic, emphasizing the intricate interplay between economic disparities and healthcare inequities. Access to vaccines and treatment, a central concern, is scrutinized in-depth, shedding light on the complex dynamics of vaccine distribution and the imperative of equitable access. We examine public health messaging, grassroots initiatives, and the pivotal roles played by governments and NGOs in mitigating the crisis. Throughout the chapter, we provide case studies from diverse African nations, offering a nuanced understanding of their unique experiences and responses to the pandemic. By distilling key lessons learned, we provide a roadmap for future healthcare policies and practices, advocating for proactive preparedness in the face of future health crises. In conclusion, "COVID-19 Lessons and Medical Inequalities in Africa" serves as an essential resource for policymakers, healthcare professionals, scholars, and advocates committed to addressing healthcare disparities in Africa and beyond. It underscores the urgency of addressing medical inequalities, not just as a

A. S. Omar (✉)
Ministry of Energy and Water Resources, Federal Government of Somalia, Mogadishu, Somalia
e-mail: dr.abdiqani@hotmail.com

B. W. K. Son (ed.), *The Landscape of Global Health Inequity*, Integrated Science 22,
https://doi.org/10.1007/978-3-031-60502-4_5

response to pandemics, but as a fundamental component of building resilient and equitable healthcare systems.

Keywords

Healthcare • Leadership • Pandemic • Management and crisis

1 Introduction

The COVID-19 pandemic, caused by SARS-CoV-2, has had a substantial impact on every part of the world, including the African continent. The virus's first appearance in Africa was traced back to Egypt in February 2020, when the first verified case was published. This person had lately returned from China, where the pandemic began. This was the virus's first appearance on the continent [1]. As the epidemic progressed, the virus spread rapidly inside Africa via various pathways, mainly owing to variables including international travel, population density, and local transmission dynamics. The contemporary world's interconnection aided the virus's rapid dissemination via air travel and other modes of transportation [2]. Because of the difficulties in establishing effective containment measures, urban areas with dense populations and inadequate access to healthcare facilities become particularly vulnerable to the virus's spread [3].

In addition, the distinctive socioeconomic environment of many African nations contributed to the spread of the virus. In certain areas, it was difficult to implement rigorous lockdown measures because of the reliance on informal economies, where many people must work every day to feed their families [4]. Additionally, it was difficult to detect and isolate individuals and provide proper medical care in some locations due to a lack of healthcare infrastructure.

It became evident that managing the pandemic's effects in Africa required a multifaceted approach since the virus propagated throughout communities, particularly during gatherings and celebrations. To lessen the effect of the virus and safeguard vulnerable people, governments and health authorities throughout the continent started a number of actions, including lockdowns, travel restrictions, testing campaigns, and public awareness campaigns.

2 COVID-19 and Healthcare Landscape in Africa

Many African nations struggled with their healthcare system's capacity to handle a surge in cases, revealing gaps in preparedness. The socio-economic impact has been substantial, with disruptions to livelihoods and economies. Public health measures, including social distancing and lockdowns, were implemented, but their effectiveness varied across regions. Vaccine distribution and hesitancy also emerged as critical issues, with concerns about equitable access and public acceptance [5]. While Africa has shown resilience and adapted strategies to combat the virus, the situation remains dynamic, and ongoing efforts are necessary for an effective response. As cited by Salyer et al. [6], researchers and policymakers continue to assess lessons learned and formulate recommendations for future public health emergencies, aiming to strengthen healthcare systems and ensure a more robust response in the face of emerging health threats.

The COVID-19 pandemic has significantly impacted the healthcare landscape in Africa, exposing vulnerabilities in the continent's healthcare infrastructure and socio-economic fabric [7]. Challenges such as limited healthcare resources, uneven access to medical facilities, and socio-economic vulnerabilities became pronounced, particularly in regions with weaker healthcare systems [8]. Disparities in healthcare infrastructure were evident, with urban areas having more resources than rural regions, exacerbating healthcare inequalities [9].

Testing capabilities varied across the continent, and underreporting in some areas made it challenging to accurately assess the full extent of the pandemic's impact [9]. Furthermore, the economic fallout, affecting sectors such as tourism, agriculture, and informal economies, exacerbated existing conditions like poverty and malnutrition, heightening the vulnerability of populations to the health crisis [8]. The pandemic's indirect effects on routine healthcare services, such as immunization programs and maternal health, strained the overall health system, further emphasizing the need for resilient healthcare systems [9]. Urban–rural disparities also underscored the importance of targeted interventions and resource allocation [10].

In response to these challenges, various African countries implemented innovative strategies, leveraging technology for contact tracing, public health communication, and monitoring the spread of the virus. Collaborative efforts within the African Union and partnerships with international organizations played a crucial role in securing medical supplies and facilitating information exchange [9]. Vaccine rollout in Africa faced hurdles of access, affordability, and hesitancy. Initiatives like COVAX aimed to provide equitable access to vaccines, but distribution challenges persisted [11]. Community engagement and addressing misinformation became paramount in overcoming vaccine hesitancy [11].

As the situation evolves, ongoing research and policy adjustments are essential to navigating the complex interplay of health, socio-economic, and cultural factors in the African context. The lessons learned from this pandemic are instrumental in building resilience and reinforcing healthcare systems to better withstand future global health challenges in the region [9]. To stay abreast of the latest developments, it is crucial to refer to real-time data, peer-reviewed articles, and reports from reputable health organizations.

3 African Responses to COVID-19

The African Union and regional organizations were critical in encouraging international collaboration, allowing for the sharing of resources, knowledge, and best practices [9]. However, difficulties in healthcare infrastructure remained, including insufficient testing capacity, medical supply shortages, and overburdened healthcare systems [9]. Some governments launched economic assistance measures, such as cash transfers and food distribution programs, to ease the socioeconomic burden on vulnerable communities [8]. Vaccine distribution efforts were hampered by access, price, and logistics issues, with vaccine hesitancy creating additional obstacles in certain areas [11].

Throughout the epidemic, African nations demonstrated adaptation and resilience by repurposing existing mechanisms for COVID-19 response and actively engaged in research and innovation [9]. The pandemic's lessons stressed the significance of strengthening healthcare systems, investing in research and development, and enhancing coordination in the face of future health emergencies [9]. The situation's dynamic nature necessitates constant attention, with governments, public health agencies, and international organizations constantly altering responses depending on changing conditions.

Several nations implemented rigorous preventative measures, like lockdowns, travel restrictions, and social distancing, in the early phases of the virus's transmission [9]. With the development of mobile applications and data analytics for contact tracing and public health communication [9], technology played a critical role. Community participation emerged as a key component of solutions, with governments, non-governmental organizations, and community leaders working together to spread accurate information and dispel myths [10].

Vaccine distribution efforts were hampered by access, price, and logistics issues, with vaccine hesitancy creating additional obstacles in certain areas [11]. In response, some African countries introduced novel vaccine distribution and uptake tactics, including mobile immunization clinics and community engagement campaigns.

Throughout the epidemic, African nations demonstrated adaptation and resilience by repurposing existing mechanisms for COVID-19 response and actively engaged in research and innovation [9]. The pandemic's lessons stressed the significance of strengthening healthcare systems, investing in research and development, and enhancing coordination in the face of future health emergencies [9]. The situation's dynamic nature necessitates constant attention, with governments, public health agencies, and international organizations constantly altering responses depending on changing conditions.

Furthermore, some African nations used traditional medicine and indigenous knowledge systems to supplement Western medical techniques, emphasizing the necessity of incorporating local practices into the larger public health response [9]. Furthermore, collaboration with the commercial sector, non-governmental organizations, and international partners was critical in mobilizing resources and knowledge to assist healthcare systems and pandemic response operations [10].

These collaborations underscored the interdependence of global health and the importance of collective effort in addressing the difficulties faced by a pandemic of this magnitude.

3.1 Healthcare and Economic Inequities

African medical inequities have been forcibly brought to light by the COVID-19 epidemic, teaching important lessons about the pressing need to overcome these differences. The significance of fair access to healthcare is one important lesson. The epidemic exposed the stark differences between urban and rural locations in healthcare infrastructure, resources, and services. While rural people in African nations struggled to acquire even the most basic healthcare services, metropolitan areas had greater access to diagnostic, therapeutic, and medical facilities [12].

The vulnerability of underprivileged communities is another crucial lesson. Vulnerable populations, such as refugees, internally displaced people, and those residing in unofficial settlements, were disproportionately impacted by the epidemic. These populations were at increased risk due to the lack of access to information, sanitation, and healthcare, highlighting the need for specialized actions to safeguard their health and well-being [13].

The epidemic also made it clear how crucial it is to boost domestic healthcare manufacturing capabilities. The dependence of African countries on foreign supply networks for medical supplies and equipment highlights their susceptibility to interruptions. To lessen their reliance on outside suppliers, some nations realized this and began investing in the domestic manufacture of necessary medical goods [3].

Numerous African countries have seen notable disparities in access to healthcare services, highlighting the persistent issue of medical inequality. One country that has suffered from severe medical differences is Nigeria. In Nigeria, there have been significant healthcare inequalities that have an impact on different demographic groups in terms of access and results. The nation's inadequate healthcare infrastructure, underfunding of the health sector, and limited access to high-quality medical care all contribute to significant variations in health outcomes [14]. These variations highlight the urgent requirement for comprehensive actions to address the root causes of medical disparity and offer equitable access to healthcare for all Nigerians.

Another African nation dealing with major medical disparities is South Africa. Access to healthcare services and facilities varies between urban and rural locations as a result of these differences. The divide between rich and disadvantaged populations is widened by the fact that urban centers have state-of-the-art healthcare facilities, whereas rural regions frequently lack adequate medical treatment [15]. The disparity between urban and rural areas has highlighted the need for efforts to close it and guarantee fair access to healthcare across the country.

In Kenya, the disparity in healthcare availability between rural and urban areas is particularly glaring. In comparison to metropolitan places, rural communities struggle to obtain medical facilities and high-quality care [16]. The ensuing differences highlight the necessity of removing regional restrictions on access to healthcare and fostering a fair allocation of healthcare resources throughout the nation.

3.2 Access to Vaccines and Treatment

During the COVID-19 epidemic in Africa, access to vaccinations and treatment was a difficult and varied problem. The continent struggled to get an adequate and timely supply of COVID-19 vaccinations, owing mostly to global vaccine distribution issues and restricted production capacity [10]. Initiatives like the COVAX program intended to solve these difficulties by providing fair and equal access to vaccinations throughout the world, however implementation issues and vaccine nationalism hampered the initiative's efficacy [7].

Economic factors also had a crucial role in influencing vaccination and treatment access in Africa. Due to budget constraints, several countries had to rely on financial help, donations, or subsidies to get vaccinations, exposing the economic gaps impacting healthcare access [8]. Vaccine hesitation has also developed as a key issue, fueled by misinformation, cultural beliefs, and historical concerns, necessitating focused public awareness campaigns and community engagement activities to address and minimize reluctance [11].

While worldwide alliances and donations from pharmaceutical corporations and international organizations played an important role in augmenting Africa's vaccination supply, vaccine availability difficulties continued. Donations were made to provide a particular percentage of vaccinations to low-income countries, especially those in Africa, but overall vaccine availability remained a serious problem [10].

The availability of COVID-19 therapies, including antiviral medicines and therapeutic approaches, differed throughout the continent. Affordability, regulatory permissions, and healthcare infrastructure all affected therapy availability and accessibility in various areas [9]. Some African nations investigated or begun initiatives to produce vaccines domestically to increase self-sufficiency and minimize reliance on external sources [10]. These initiatives frequently included knowledge transfer agreements and collaborations with multinational vaccine makers.

Collaboration with the commercial sector, notably pharmaceutical firms, aided in the availability of vaccinations and treatments. Partnerships included discussions on price, distribution, and local manufacturing arrangements, emphasizing the importance of public–private partnerships in tackling healthcare difficulties during the pandemic [9]. The fluid nature of these difficulties highlights the continuous efforts required to address inequality, improve vaccine delivery methods, and increase treatment access to achieve a complete and equitable response to Africa's ongoing pandemic.

3.3 Roles of Governments and NGOs

Governments and non-governmental organizations (NGOs) played critical roles in reacting to the COVID-19 epidemic in African nations, establishing public health measures, providing healthcare services, and addressing socioeconomic difficulties. Here are some instances of government and non-governmental organization (NGO) roles:

Governments' Roles:

1. **Policy Development and Implementation**: Governments have been at the forefront of developing and enforcing measures to combat the spread of COVID-19. This involves imposing lockdowns, limiting travel, and enforcing social distancing measures. Many African countries have attempted to boost healthcare infrastructure by raising testing capabilities, establishing isolation centers, and improving medical facilities. South Africa, for example, increased diagnostic capacities and constructed field hospitals to deal with the influx of cases.
2. **Economic Assistance Measures**: Governments have introduced economic assistance measures to help disadvantaged communities cope with the epidemic. In Nigeria, for example, the government launched cash transfer programs and food distribution to aid individuals affected by lockdowns.
3. **Vaccine Rollout Programs**: Governments are driving vaccine distribution and administration activities. Ethiopia, for example, started a countrywide COVID-19 immunization program, focusing on healthcare personnel and vulnerable communities.
4. **Public Communication and Awareness**: Governments have been in charge of public communication and awareness efforts to spread proper information regarding COVID-19. Kenya's government used a variety of communication platforms to educate people on preventive measures.

Roles of NGOs:

1. **Community Outreach and Education**: Non-governmental organizations (NGOs) have played an important role in community outreach and education, offering information on COVID-19 and preventative measures. Amref Health Africa, for example, ran community-based awareness programs in a number of nations.
2. **Healthcare Provision and Support**: Non-governmental organizations (NGOs) have assisted in the provision of healthcare services by giving medical supplies, personal protective equipment (PPE), and aiding in the creation of testing and treatment facilities. Médecins Sans Frontières (MSF) has been providing medical treatment and assistance in a number of African nations.
3. **Support for Vulnerable Populations**: NGOs have prioritized helping vulnerable groups such as refugees and internally displaced people. The International

Rescue Committee (IRC) has assisted displaced communities in nations such as South Sudan with healthcare and sanitation.

4. **Policy Participation and Advocacy**: NGOs have engaged in lobbying attempts to influence COVID-19 response policy. Organizations such as the Zimbabwe Association of Doctors for Human Rights fought for better healthcare infrastructure and working conditions for healthcare workers in Zimbabwe.

4 COVID-19 Lessons in Africa

The COVID-19 pandemic has taught Africa valuable lessons, altering the continent's approach to public health disasters. The urgency to build healthcare infrastructure, highlighting the need for robust testing capacities, well-equipped medical facilities, and professional healthcare personnel, is foremost among these lessons [10]. Effective community participation and communication are identified as essential components, emphasizing the significance of establishing trust and eliminating misconceptions [9]. Africa's resilience and inventiveness were also emphasized by the epidemic, as governments recycled infrastructure and used technology for contact tracking [9].

Global collaboration emerged as an important lesson, underlining the need for collaborations with international organizations and adjacent nations [10]. Economic resilience has taken precedence, forcing a rethinking of diverse and resilient economic policies [17, 18]. The difficulties in vaccine distribution highlighted the need for local vaccine production, spurring initiatives to establish manufacturing capabilities on the continent [11]. As crucial lessons, improved data and surveillance systems, investment in research and development, cross-sector cooperation, and a focus on readiness and early reaction were noted [9]. These findings provide a basis for strengthening health systems and improving tactics to effectively manage future public health concerns in Africa.

The African COVID-19 pandemic has shown the crucial need for cross-sectoral teamwork in responding to health disasters. The interdependence of public health, education, and the private sector became clear, underlining the importance of a multidisciplinary approach [9]. Lessons learned highlighted the need for collaboration outside the field of healthcare. Partnerships with the commercial sector, for example, allowed access to resources and experience, which was critical in addressing the pandemic's numerous issues [9]. This comprehensive approach provides a thorough grasp of the issues at hand, as well as the development of inclusive initiatives that take into consideration the socioeconomic and cultural complexities of African nations. Moving forward, the emphasis on cross-sectoral partnerships is a critical lesson for developing resilient systems capable of handling the complex dynamics of future public health emergencies in the area [19].

5 Conclusion

Through comprehensive examination, it is evident that governments and NGOs played crucial roles in responding to the COVID-19 pandemic in African nations. The knowledge gained from these medical disparities has generated debates and initiatives to increase healthcare accessibility and inclusiveness. Governments, international organizations, and civil society have started initiatives to close healthcare service gaps, improve healthcare infrastructure in neglected regions, and guarantee that vulnerable populations are given the attention they need during health emergencies [20]. The foundation for more extensive, egalitarian, and resilient healthcare systems in Africa has been set by these lessons. In addition to helping to manage the current crisis, these lessons will probably have an impact on future policy and practice. Due to the lessons learned, adaptations made, and a newfound emphasis on community involvement, utilization of local expertise, and the establishment of an adaptive government, Africa is now better equipped to handle forthcoming crises. The pandemic reinforced healthcare systems and hastened the deployment of digital solutions. Even while difficulties still exist, COVID-19's lessons have inspired solutions and partnerships that will strengthen the continent's long-term resilience.

References

1. BBC News (2020a) Coronavirus: Egypt confirms first case, says affected person is foreigner. https://www.bbc.com/news/world-middle-east-51486173. Accessed 10 Aug 2022
2. WHO (2020a) Senegal steps up fight against COVID-19. https://www.afro.who.int/news/senegal-steps-fight-against-covid-19. Accessed 4 Apr 2020
3. African Union (2020) Africa and COVID-19: a comprehensive response to the pandemic. https://au.int/en/africacdc/covid19/resources. Accessed 3 Jan 2024
4. PLoS Neglected Tropical Diseases (2021) COVID-19 in Africa: diverse responses to a common threat. https://journals.plos.org/plosntds/article?id=10.1371/journal.pntd.0009496 Accessed 19 Dec 2021
5. Gilbert M, Pullano G, Pinotti F, Valdano E, Poletto C, Boëlle PY, Colizza V (2020) Preparedness and vulnerability of African countries against importations of COVID-19: a modelling study. The Lancet 395(10227):871–877
6. Salyer SJ, Maeda J, Sembuche S, Kebede Y, Tshangela A, Nkengasong J (2021) The first and second waves of the COVID-19 pandemic in Africa: a cross-sectional study. The Lancet 397(10281):1265–1275
7. WHO (2020b) Coronavirus disease (COVID-19) pandemic. https://www.who.int/emergencies/disease-outbreak-news/item/2020-DON264. Accessed 6 May 2020
8. World Bank (2021) Africa's Pulse, No. 23, October 2021: an analysis of issues shaping Africa's economic future. https://www.worldbank.org/en/region/afr/publication/africas-pulse. Accessed 9 Oct 2021
9. Adeyemo AO, Aborode AT, Bello MA, Obianuju AF, Hasan MM, Kehinde DO, Hossain MS, Bardhan M, Imisioluwa JO, Akintola AA (2022) Malaria vaccine: the lasting solution to malaria burden in Africa. Ann Med Surg 79:104031. https://doi.org/10.1016/j.amsu.2022.104031
10. WHO Africa (2020) COVID-19 in Africa: protecting lives and economies. https://www.afro.who.int/publications/covid-19-africa-protecting-lives-and-economies. Accessed 15 Apr 2020

11. Nkengasong JN, Mankoula W (2020) Looming threat of COVID-19 infection in Africa: act collectively, and fast. Lancet 395(10227):841–842
12. UNDP (2020a) COVID-19 socioeconomic impact assessment: Ghana. https://www.gh.undp.org/content/ghana/en/home/library/crisis_prevention_and_recovery/socio-economic-impact-assessment-of-covid-19-in-ghana.html. Accessed 7 June 2021
13. Crisp J (2020) UNHCR at 70: an uncertain future for the international refugee regime. Glob Gov 26(3):359–368
14. Oyediran KA, Isiugo-Abanihe UC, Uzochukwu BSC (2018) Inequities in health care services utilization in Nigeria. Int J Equity Health 17(1):39. https://doi.org/10.1186/s12939-018-0754-5
15. Hall K, Durrheim K, Wademan DA, Mathews C (2017) Poor people's health care: health equity and urban areas in South Africa. Int J Public Health 62(Suppl 1):135–140. https://doi.org/10.1007/s00038-017-1049-9
16. Nzinga J, Mbindyo P, Mbaabu L, English M (2009) Service delivery in Kenyan district hospitals—what can we learn from literature on mid-level managers' roles? Hum Resour Health 7(1):69. https://doi.org/10.1186/1478-4491-7-69
17. World Bank (2021) Africa's Pulse, No. 23, October 2021: an analysis of issues shaping Africa's economic future. https://www.worldbank.org/en/region/afr/publication/africas-pulse. Accessed 8 Oct 2021
18. African Development Bank (2020) African economic outlook 2020 supplement: amid COVID-19. https://www.afdb.org/en/knowledge/publications/african-economic-outlook. Accessed 30 Jan 2020
19. UNDP (2020b) Ethiopia: building a resilient health system during COVID-19. https://www.africa.undp.org/content/rba/en/home/stories/ethiopia--building-a-resilient-health-system-during-covid-19.html. Accessed 1 June 2022
20. Ogunleye O, Basu D, Mueller D, Sneddon J, Seaton RA, Yinka-Ogunleye AF, Godman B (2020) Response to the novel coronavirus (COVID-19) pandemic across Africa: successes, challenges, and implications for the future. Front Pharmacology 1205

A Review of Global Inequities in COVID-19 Vaccination Access and Uptake

6

Alessandro Siani

Abstract

The rapid development of safe and effective vaccines against COVID-19 is aptly considered a watershed moment in the course of the pandemic, and one of the greatest collaborative accomplishments of modern medicine. However, the rollout of vaccination campaigns highlighted striking global inequities in the procurement and distribution of vaccine doses. Low- and medium-income countries faced significant struggles hindering both the domestic production of vaccines and their acquisition from global suppliers. The insufficient access to vaccine supplies in developing countries was caused by a combination of external (vaccine nationalism, intellectual property rights, unfair competition) and internal (low economic means, limited manufacturing facilities, gaps in technical expertise) factors. In addition to the procurement and manufacturing hurdles, the administration of the limited available stocks was hindered by both logistical (inadequate supply chain, transportation, and healthcare infrastructure) and sociocultural (vaccine hesitancy, scientific misinformation, religious opposition) issues. Despite the roll-out of multiple international initiatives to facilitate global access to COVID-19 vaccinations, the inequities in vaccine distribution resulted in limited and delayed vaccine coverage—and ultimately immeasurable loss of health and life—in the global South. This chapter discusses the socioeconomic, geopolitical, and cultural causes of COVID-19 vaccine inequity, its impact on the affected populations, and the effectiveness of global initiatives designed to mitigate it.

A. Siani (✉)
School of Biological Sciences, University of Portsmouth, King Henry Building, King Henry 1st Street, Portsmouth PO1 2DY, UK
e-mail: alessandro.siani@port.ac.uk

B. W. K. Son (ed.), *The Landscape of Global Health Inequity*, Integrated Science 22,
https://doi.org/10.1007/978-3-031-60502-4_6

Keywords

COVID-19 vaccination • Vaccine distribution • Global health inequities • Economic disparities • LMIC • TRIPS waiver • COVAX • Vaccine apartheid

1 Introduction

The COVID-19 pandemic has reinforced the crucial importance of vaccination campaigns in controlling the spread of infectious diseases and minimizing the associated loss of life and wellbeing. While safe and effective vaccines against SARS-CoV-2 were developed with exceptional speed thanks to funding and collaborative efforts unprecedented in human history [1], global vaccine rollout has proceeded with stark disparities in access and uptake between developed and developing countries.

This chapter seeks to evaluate the disparities in global COVID-19 vaccination coverage, highlighting its impact and the underpinning social, economic, cultural, and political causative factors. It also accounts for the effectiveness of initiatives aimed at facilitating global access to COVID-19 vaccinations and outlines the lessons learned from the COVID-19 pandemic in terms of the causes, effects, and mitigation of global health inequities. It is essential to highlight that inequities in access to healthcare and vaccination occur at both the intranational and international levels. While this chapter focuses largely on global disparities between High-Income Countries (HICs) and Low- and Medium-Income Countries (LMICs), it is essential to acknowledge that intranational healthcare inequities have a dramatic impact on the affected populations even in affluent countries, and their understanding contributes to our ability to recognize and address global challenges.

In a globalized world, the ramifications of inconsistent vaccine distribution transcend national borders: the high infection rates caused by suboptimal vaccine coverage in an area increase the likelihood of viral mutation [2]. This, in turn, has the potential to result in the emergence of new viral variants with increased infectiousness and the ability to evade immune detection, which would rapidly circulate and cause further outbreaks across the planet.

The term "inequity" is used throughout this chapter to signify, in accordance to the Oxford English Dictionary, "Want of equity or justice; the fact or quality of being unfair; unfairness, partiality". It is important to note that although the word "inequality" is defined as "The condition or fact of having different rights, advantages, or opportunities" without implying a moral judgment regarding the unfairness of such condition, a more modern definition clarifies that the term is frequently used "esp. according to differences in gender, ethnicity…". Therefore, while some authors cited in this chapter use the terms interchangeably, others may express a preference for either term depending on whether they are referring to a factual state of affairs (inequality) or highlighting the unfairness of the situation (inequity).

As will be discussed in the next section, the analysis of global vaccination inequities is complicated by the disparity in healthcare data collection and reporting in different parts of the world, which confounds the comparison of trends between different countries and has the potential to configure a paradoxical "unknown unknowns" scenario [3].

2 Global Disparities in Diagnosis and Data Collection

The evaluation of the global incidence and impact of COVID-19 is confounded by wide disparities in the extent to which the disease is diagnosed and treated in different regions of the world. At the onset of the pandemic, the scarcity of diagnostic tests across the global market resulted in their uneven distribution, caused by unfair competition that favored countries with the highest economic means [4]. Insufficient access to diagnostic testing in low- and middle-income countries was caused by multiple factors. Compared to high-income countries, LMICs often lack the industrial infrastructure required to manufacture diagnostic tests, leaving them to rely on imports from global suppliers [5]. However, as previously mentioned, the competition for a limited supply of testing equipment inevitably favours countries with higher economic means. Moreover, LMICs tend to have lower availability of testing laboratories, meaning that they have a reduced capacity to carry out lab-based tests (e.g. RT-PCR or ELISA), and they have to heavily rely on rapid antigen tests (e.g. LFT). Despite their unquestionable diagnostic utility, rapid antigen tests are typically less sensitive and accurate than molecular tests, and their results are often not communicated to health authorities by their users [6].

The reliance on rapid antigen testing and the low availability of laboratory facilities in LMICs also impact their ability to sequence viral genomes. Genomic sequencing of SARS-CoV-2 is fundamental to the efforts of diagnosing and treating COVID-19 and developing vaccination campaigns. A recent study revealed that while 78% of HICs were able to sequence at least 0.5% of their COVID-19 cases, this was only the case for 42% of LMICs [7].

The inadequate collection of data on COVID-19 incidence and the low rate of genomic sequencing has a detrimental effect on the ability of LMICs to design and deploy vaccination campaigns, compounding the numerous other factors underpinning global inequities in vaccine distribution and access. The evaluation of the extent and impact of global health inequities is further confounded by the lack of a standardized approach to public health modelling, hampering the comparability of outcomes between studies that use different modelling strategies.

3 Impact of the Global Inequities in Vaccine Distribution and Access

As of the writing of this chapter in January 2024, a total of 13.53 billion COVID-19 vaccine doses have been administered globally, and 70.6% of the world population has received at least one dose to date [8]. However, according to the same source, the percentage of people who have received at least one dose ranges from 32.9% in low-income countries (LICs) to 79.9% in HICs. The global disparity in vaccine coverage is even more accentuated with respect to the administration of booster doses (defined as "doses administered beyond those prescribed by the original vaccination protocol"), with booster coverage ranging from 4.1% in LICs to 66.4% in HICs.

A data-driven, age-stratified epidemic modelling analysis investigated the impact of COVID-19 vaccine inequities in countries with different income levels [9]. The study revealed that between 54 and 94% of COVID-19 deaths in the analyzed LMICs could have been avoided if vaccination rates had been on par with HICs. Interestingly, even in counterfactual scenarios where LMICs had comparable early access to vaccine doses as HICs, a significant percentage (6–50%) of avoidable deaths was still observed in LMICs. This observation indicates that the availability of vaccine doses is only one of the factors underpinning the excess mortality associated with COVID-19 in LMICs. Based on the results of the modelling analysis, the authors concluded that "vaccine inequity, in both the number of doses available and the timeline of delivery, drastically reduced the impact of vaccination campaigns in the group of LMIC studied".

The interplay between income levels and the timing of access to COVID-19 vaccines in different countries was also investigated by Duroseau and colleagues via an ecological study [10]. The investigation revealed that "Countries with the smallest economies reported first vaccination much later than larger economies on all three rankings, as much as 100 days longer". Moreover, every single day the first vaccination was delayed resulting in additional increases in cumulative COVID-19 cases and mortality.

Global epidemiological data indicates that, after one year since COVID-19 vaccines became available, vaccine coverage ranged from 0.03% in Burundi to 98.99% in the United Arab Emirates [11]. Regression analysis revealed that vaccine coverage was strongly associated with a country's Human Development Index (HDI). A significant reduction in viral reproduction rate, and consequently COVID-19 cases, hospitalizations, and deaths was observed in countries (typically HICs) where vaccine coverage was above 70%.

It is crucial to note that inequities associated with the response to the pandemic did not just affect the access to COVID-19 vaccinations, but also the uptake of routine childhood immunization. A systematic literature review identified 17 factors disrupting access to and uptake of childhood vaccinations [12]. The study highlighted that 80% of those factors were related to healthcare-seeking behaviours, particularly avoidance of healthcare settings and fear of contracting COVID-19. According to recent studies, the key factors underpinning vaccine inequity include "vaccine nationalism, intellectual property rights, constraints in manufacturing capacity, poor resilience of healthcare systems, and vaccine hesitancy" [13], and "the limited ability of LMICs to purchase significant quantities of the vaccine" [14]. The nature and impact of the key causative factors of low vaccine access in LMICs will be evaluated in more detail in the following sections.

4 Economic and Geopolitical Factors Influencing Vaccination Access

According to global management consulting firm McKinsey & Company, the effective rollout of vaccines must meet the following critical considerations: the vaccine has to be "available, administrable, accessible, acceptable, affordable, accountable" [15]. The availability of vaccines in LMIC was constrained by their lower financial means, which limited their ability to purchase vaccine stocks from global suppliers. The acquisition of vaccine stocks by LMICs was further complicated by the phenomenon known as "vaccine nationalism", which has been defined as "the mindset and act of gaining preferential access to newly developed COVID-19 vaccines by individual countries (higher-income countries, in practice)" [16]. It has been suggested that vaccine nationalism is associated with multiple negative consequences for global public health, including (1) increased competition between global superpowers to develop vaccines which might lead to forcibly speeding up clinical trials, potentially compromising their safety and effectiveness, (2) a disruption of global supply chains due to hoarding of key components and unwillingness to cooperate, and (3) preferential deals between vaccine supplier and wealthier countries, leading to uneven and unfair global distribution of vaccine stocks [17].

In addition to the financial and political hurdles limiting the purchase of vaccine stocks from foreign suppliers, LMICs were severely limited in their ability to produce vaccines locally due to a combination of intellectual property (IP) rights and constraints in manufacturing capacity [13]. At the height of the pandemic, it was noted that "Ending this global pandemic seems unlikely with the current patent and technology sharing systems because (i) the number of available vaccine doses is limited by the production capacity of the companies owning the patents, recipes, and technology and (ii) the inequitable distribution of the limited supply of vaccines throughout the world" [18]. While several pharmaceutical companies based in LMICs had the expertise and technical means required to produce COVID-19 vaccines, this could only be achieved if the patent owners agreed to waive their intellectual property rights, share the vaccine recipes, and train local workforce in the preparation procedure. In October 2020, India and South Africa proposed to the World Trade Organization (WTO) that the agreement on Trade-Related Aspects of Intellectual Property Rights (TRIPS) be temporarily waived with respect to COVID-19-related medical technologies, including therapeutics, diagnostics and vaccines [19]. While most WTO members supported the proposal, several HICs and pharmaceutical companies opposed the TRIPS waiver, leading to lengthy negotiations. The TRIPS waiver was ultimately approved in June 2022, however, the negotiations resulted in a 20-month delay during which LMICs remained largely unable to manufacture their own vaccine doses. Moreover, the approved TRIPS waiver was heavily criticized for being "watered down" [20] and "considerably narrower in scope than the original proposal" [21].

In addition to IP issues, many LMICs face considerable obstacles in the domestic production of vaccines and "have not been able to establish their own manufacturing units because of their human, financial and technical resource constraints in vaccine manufacturing" [22]. A survey of 26 vaccine-manufacturing companies based in developing countries was carried out to investigate the barriers to domestic vaccine production [23]. The study revealed that "developing country vaccine manufacturers have diverse capabilities to develop, produce, and supply vaccines for local and global use", however "regulatory challenges and price competition are key impediments to manufacturer viability".

While vaccine availability played an undeniable role in contributing to low vaccine coverage in LMICs, it has been argued that "the bulk of attention and support has been focused on vaccine access; ultimately, the bottleneck to coverage may come from in-country delivery", highlighting the crucial importance of internal logistics in vaccine distribution [15]. In a recent study, a panel of global experts on vaccine delivery and implementation was interviewed to investigate the challenges hindering the implementation of COVID-19 vaccination programmes in LMICs [24]. Thematic content analysis of their responses indicated that "challenges to vaccine program implementation include issues related to equity; governance, decision-making, and financing; regulatory structures, planning, and coordination; prioritisation, demand generation, and communication; vaccine, cold chain, logistics, and infrastructure; service delivery, human resources, and supplies; surveillance, monitoring, and evaluation". Similarly, an interview of national-level stakeholders (policymakers and immunisation programme experts) from Ghana and Bangladesh revealed that barriers to vaccine distribution in their respective countries could be broadly classified in supply-side barriers and demand-side barriers [25]. In addition to the hurdles to purchasing and manufacturing vaccines that were discussed earlier in this chapter, supply-side barriers include limitations in cold-chain capability (e.g. limited availability of refrigerators, inconsistent power supply, poor roads), administrative issues related to the provision and delivery of vaccination services, and challenges associated with program surveillance and data management. Demand-side barriers include cultural, religious, and social factors typically associated with vaccine hesitancy, which will be discussed in the next section.

5 Vaccine Hesitancy, Cultural Influence and Misinformation

The mistrust, delay or refusal of vaccinations, encompassed by the umbrella term "vaccine hesitancy", is a centuries-old phenomenon predating both the COVID-19 pandemic and Andrew Wakefield's MMR scandal [26]. Over the years, the World Health Organization (WHO) has produced frameworks (most notably, the "3Cs" and the "Matrix of Determinants") to categorise the causes underpinning vaccine hesitancy and provide guidance to governments and healthcare workers in their efforts of tackling it [27]. A comparison of cross-sectional studies carried out in 2019 and 2022 revealed that vaccine confidence (one of the three Cs in the 3Cs model) declined considerably since the onset of the COVID-19 pandemic [28]. More recently, a UNICEF report corroborated this observation, warning that during the pandemic confidence in childhood vaccinations declined by more than a third in 52 out of 55 countries studied, including both HICs and LMICs [29].

In July 2021, a Nature Medicine study analysing survey data from 44,260 participants in 10 LMICs and the United States of America revealed that vaccine acceptance appeared to be considerably higher (mean 80%) in LMICs compared to the USA (64%) [30]. Interestingly, a March 2022 meta-analysis of 36 studies investigating vaccination attitudes in 83,867 survey respondents across 33 LMICs revealed that COVID-19 vaccine acceptance in the study population was on average 59%, ranging from 43% in Egypt to 77% in India [31]. While the wide disparity in vaccine acceptance in LMICs as measured in the two studies might indeed be explained by differences in methodological approach and sample population, the lower vaccine acceptance reported in the 2022 study might be at least partially attributable to the detrimental impact of the pandemic on vaccine confidence. This discrepancy also reinforces the observation that the inconsistent availability and reliability of healthcare-related data complicate the evaluation of the causes and consequences of vaccine inequity.

Data collected in the largest global health survey to date, involving over 14 million participants across 90 countries, were analysed to investigate the role of education inequality in COVID-19 vaccine acceptance [32]. Interestingly, the study revealed that "Education-related inequality in structural barriers, vaccine hesitancy, and vaccine refusal varied across countries, and was most pronounced in high-income countries, overall. Low-income and lower-middle-income countries reported widespread experiences of structural barriers and high levels of vaccine hesitancy alongside low levels of education-related inequality". These findings corroborate previous observations alluding to a non-linear relationship between education levels and vaccine hesitancy within a population [33], and highlight how the extent (and, in some cases, direction) of the relationship varies between countries with different income levels.

Religious beliefs are a key contributor to people's attitudes and behaviours towards vaccination across virtually all countries and religions [34]. However, it has been observed that there are no consistent global patterns in the interplay between religion and vaccine acceptance/hesitancy, as religious beliefs may have positive or negative effects on vaccine attitudes depending on the specific context of each region and population [35]. For example, while religious fundamentalism has been identified as a major causative factor of polio vaccine refusal within Muslim populations in Afghanistan, Nigeria and Pakistan, religious objection rate is very low (2%) in Saudi Arabia—a country where Muslims represent nearly 100% of the population. Likewise, while strong opposition to COVID-19 vaccinations has been reported among some Christian communities, other Christian organisations "place far greater store on science as a gift from God, medical science as a means of transforming societies for good, and the potential of vaccines to control a rampant pandemic" [36]. Religious organizations played a key role in COVID-19 crisis response and management within vulnerable and marginalized communities, as exemplified by a study evaluating the work of faith-based organizations of Catholic, Islamic, and Buddhist inspiration based respectively in Brazil, Indonesia, and Sri Lanka [37].

Misinformation about the safety and efficacy of the COVID-19 vaccines is a major determinant of vaccine hesitancy and refusal [38, 39]. A mixed-methods analysis of misinformation related to COVID-19 vaccines revealed that "people were exposed to conspiratorial misinformation about COVID-19 vaccines such as the vaccines including a microchip or them being dangerous and harmful, causing death, or altering DNA" [40]. In addition to the factors that had characterised opposition against previous vaccination campaigns, controversy and misinformation surrounding COVID-19 vaccines were further stoked by concerns related to their rapid development and by the polarisation of the political discourse around mandatory vaccination. Survey responses from 1257 participants in 35 countries revealed a strong negative correlation between a country's Gross Domestic Product and the prevalence of misinformation, whereby misinformation was significantly more frequent in poorer countries [41]. According to the authors of the study, these findings "imply that the potential harm of misinformation could be more substantial for low-income countries than high-income countries. Countries with poor infrastructures might have to combat not only the spreading pandemic but also the COVID-19 infodemic, which can derail efforts in saving lives". In a different study, a three-pronged approach was proposed to combat the COVID-19 infodemic in LMICs [42]. This approach involves a combination of pre-emptive measures (e.g. clearly defining the threat posed by misinformation and educating the public in recognising using a clear and accessible language), immediate measures (e.g. monitoring the "outbreaks" of misinformation and working in partnership with social media corporations to curb them) and long-term measures (e.g. promoting long-term health literacy by embedding infodemiology in school curricula and training community healthcare workers on fact-checking). Likewise, a review of the strategies to foster vaccine acceptance in LMICs suggested that "direct engagement of community mobilisers and frontline workers with the community through consultations, faith leaders and religious meetings, in simple non-medical terms helped increase vaccine acceptance"[43].

6 Global Initiatives and Collaborations

Since the onset of the pandemic, multiple international initiatives have been launched to tackle global inequities in the access to diagnostic, preventive and therapeutic means to respond to the COVID-19 crisis. In September 2020, the WHO launched a global partnership to provide LMICs with 120 million COVID-19 diagnostic tests [44]. This initiative was part of the Access to COVID-19 Tools (ACT) Accelerator, a collaborative partnership of 10 health and development agencies founded "to accelerate the development, production, and equitable access to COVID-19 tests, treatments, and vaccines" [45]. One of the pillars of the ACT Accelerator was the COVID-19 Vaccines Global Access (COVAX) initiative, which distributed vaccines to 92 LMICs for a total of approximately 2 billion doses, estimated to have prevented around 2.7 million deaths. COVAX was built on the public–private partnership (PPP) model, which has been criticised for "giving pharmaceutical corporations substantial power and making public representation, transparency, and accountability elusive" [46]. A scoping review published in 2023 sought to evaluate the implementation challenges and framework gaps of the COVAX initiative, and investigate the underlying governance shortcomings [47]. The review identified several implementation challenges that affected COVAX, including vaccine nationalism and export restrictions by HICs, lack of transparency over the deals with pharmaceutical companies, insufficient technology transfer and manufacturing capacity in the Global South. The study also unearthed significant global health governance and policy gaps undermining the delivery of its targets, concluding that "COVAX was left alone without adequate funding and participation and was reduced to another competitor in the vaccine market".

COVAX was not the only initiative developed to facilitate global COVID-19 vaccine access: large contributions to vaccine coverage in LMICs are attributable to bilateral or multilateral agreements between countries and to the African Vaccine Acquisition Trust (AVAT). According to a recent study, "In LMICs, 56.4% of total vaccines procured were via bilateral/multilateral agreements, 21.4% by COVAX, 4.2% by donations, and 0.5% by AVAT. COVAX delivered 1 billion doses by January 2022 which constituted 53.2% and 21.4% of procured doses in LICs and LMICs" [48].

In addition to international consortia and bilateral/multilateral agreements, vaccine access and uptake in LMICs were also facilitated by non-governmental organizations (NGOs) and grassroots initiatives. A large-scale multinational study investigated the role of civil society organizations (CSOs) in the implementation of COVID-19 response programs in LMICs [49]. CSOs are defined as "a wide array of organizations including community groups, NGOs, labor unions, social movements, as well as faith-based organizations". The study revealed that CSOs played a critical role in disadvantaged regions where governments and healthcare authorities were struggling to deliver essential resources and aid vulnerable populations during the COVID-19 crisis. The authors concluded that "CSOs across all five countries examined in the Global South adapted their missions and fulfilled a wide range of gap-filling responsibilities that can be primarily categorized in five themes: food security and livelihood support, public health and medical care, cash transfer programs, risk communication, and community education, needs assessment".

7 Conclusions

The mantra "No one is safe until everyone is safe" has been widely and appropriately used with regard to global access to COVID-19 vaccinations, including as the headline of the COVAX page on the WHO website [50]. However, while COVAX and other similar initiatives have undeniably contributed to improving vaccine coverage and reducing global inequities, the Global South remains at a stark disadvantage in terms of the extent and timing of COVID-19 vaccine access and uptake. The marginalization of LMICs with respect to access to COVID-19 vaccinations has been aptly likened to a "Vaccine apartheid" in the sense that it "divided the world according to income into countries with widespread access and those without" [51, 52]. While LMICs suffer the most direct and catastrophic consequences of vaccine inequity, its impact extends beyond the borders of the countries directly affected by it: inadequate vaccine coverage in some parts of the world might have global repercussions including emergence of new pathological agents, economic recession, supply chain disruptions and persistent unsafe travelling [13, 52]. It has been estimated that vaccine hoarding and inequitable COVID-19 vaccine coverage might have caused at least 295.8 million infections and 1.3 million deaths worldwide [53, 54]. The analysis of global access to COVID-19 vaccines reveal inequities that are reflective of broader and systematic structural violence, "in which the political economy is structured to improve and lengthen the lives of those in the Global North while neglecting and shortening the lives of those in the Global South" [55].

So, what lessons can be learned from the injustices associated with the COVID-19 response and what are humanity's prospects for change in the face of future global health crises? The delays and limitations in the implementation of COVAX clearly indicate that there can be no equitable response when global cooperation is perceived as being at odds with national interests and economic gains. In a globally interconnected world, addressing worldwide health disparities is not just a matter of human empathy, but also a key strategic imperative for our collective health and wellbeing as a species. An efficient and equitable pandemic response requires the concerted efforts of national and international stakeholders (both in the private and public sector), as well as transparency and accountability in global health governance. This endeavour would require a drastic shift in socioeconomic and political paradigms: as remarked by Gleeson and colleagues, "To enable equitable access to health products and technologies in future public health emergencies, it will be necessary to disrupt the current neoliberal paradigm and power asymmetries, through reforming the ways in which pharmaceutical R&D are funded, preventing exclusive monopolies on health products and technologies, righting the power imbalances between public and private actors and between rich and poor countries—and the ways these are reflected and reproduced in global governance structures" [55].

References

1. World Health Organization (2020) Coronavirus disease (COVID-19): vaccine research and development
2. Fan G, Song H, Yip S, He D, Zhang T (2022) Impact of low vaccine coverage on the resurgence of COVID-19 in Central and Eastern Europe. One Health 14
3. Luo J (2021) Forecasting COVID-19 pandemic: unknown unknowns and predictive monitoring. Technol Forecast Soc Change 166
4. Batista C, Hotez P, Amor YB, Kim JH, Kaslow D, Lall B, et al (2022) The silent and dangerous inequity around access to COVID-19 testing: a call to action. EClinicalMedicine 43
5. Peplow M (2020) Developing countries face diagnostic challenges as the COVID-19 pandemic surges. Chem Eng News
6. Budd J, Miller BS, Weckman NE, Cherkaoui D, Huang D, Decruz AT et al (2023) Lateral flow test engineering and lessons learned from COVID-19. Nat Rev Bioeng 1(1):13–31
7. Brito AF, Semenova E, Dudas G, Hassler GW, Kalinich CC, Kraemer MU et al (2022) Global disparities in SARS-CoV-2 genomic surveillance. Nat Commun 13(1):7003
8. Our World in Data (2024) Coronavirus (COVID-19) vaccinations. Available from: https://ourworldindata.org/covid-vaccinations
9. Gozzi N, Chinazzi M, Perra N, Vespignani A, Dean NE, Longini Jr IM et al (2023) Estimating the impact of COVID-19 vaccine inequities: a modeling study. Nat Commun 14(1)
10. Duroseau B, Kipshidze N, Limaye RJ (2023) The impact of delayed access to COVID-19 vaccines in low- and lower-middle-income countries. Front Public Health 10
11. Ning C, Wang H, Wu J, Chen Q, Gao H, Pei H (2022) The COVID-19 vaccination and vaccine inequity worldwide: an empirical study based on global data. Int J Environ Res Public Health 19(9)
12. Cardoso Pinto AM, Shariq S, Ranasinghe L, SundarBudhathoki S, Skirrow H, Whittaker E et al (2023) Reasons for reductions in routine childhood immunisation uptake during the COVID-19 pandemic in low- and middle-income countries: a systematic review. PLOS Global Public Health 3(1):e0001415
13. Ferranna M (2023) Causes and costs of global COVID-19 vaccine inequity. Semin Immunopathol
14. Sansone P, Giaccari LG, Aurilio C, Coppolino F, Passavanti MB, Pota V et al (2021) COVID-19 in low-and middle-income countries (LMICs): a narrative review from prevention to vaccination strategy. Vaccines 9(12):1477
15. Hall S, Kaplow L, Sun YS, Holt TZ (2021) 'None are safe until all are safe': COVID-19 vaccine rollout in low-and middle-income countries. McKinsey and Company
16. Zhou YR (2022) Vaccine nationalism: contested relationships between COVID-19 and globalization. Globalizations 19(3):450–465
17. Hafner M, Yerushalmi E, Fays C, Dufresne E, Van Stolk C (2022) COVID-19 and the cost of vaccine nationalism. Rand Health Q 9(4):1
18. Altindis E (2022) Inequitable COVID-19 vaccine distribution and the intellectual property rights prolong the pandemic. Expert Rev Vaccines 21(4):427–430
19. Kohler J, Wong A, Tailor L (2022) Improving access to COVID-19 vaccines: an analysis of TRIPS waiver discourse among WTO members, civil society organizations, and pharmaceutical industry stakeholders. Health Hum Rights J 24(2):159–176

20. Amin T, Kesselheim AS (2022) A global intellectual property waiver is still needed to address the inequities of COVID-19 and future pandemic preparedness. Inquiry (United States) 59
21. Rutschman AS (2023) Ministerial decision on the trips agreement (WTO). Int Leg Mater 62(2):289–294
22. Khan MI, Ikram A, Hamza HB (2021) Vaccine manufacturing capacity in low- and middle-income countries. Bull World Health Organ 99(7):479–A
23. Hayman B, Kumar Suri R, Downham M (2022) Sustainable vaccine manufacturing in low- and middle-income countries. Vaccine 40(50):7288–7304
24. Haldane V, Ariyarajah A, Berry I, Loutet M, Salamanca-Buentello F, Upshur REG (2023) Global inequity creates local insufficiency: a qualitative study of COVID-19 vaccine implementation challenges in low-and-middle-income countries. PLoS ONE 17(2):1–15
25. Tagoe ET, Sheikh N, Morton A, Megiddo I, Nonvignon J, Sarker AR et al (2021) COVID-19 vaccination in lower-middle income countries: national stakeholder views on challenges, barriers, and potential solutions. Front Public Health 9
26. Siani A (2022) Vaccine hesitancy and refusal: history, causes, mitigation strategies. In: Rezaei N (ed) Integrated science of global epidemics. Integrated science. Springer
27. World Health Organization (2014) Report of the SAGE working group on vaccine hesitancy. Available from: http://www.who.int/immunization/sage/meetings/2014/october/SAGE_working_group_revised_report_vaccine_hesitancy.pdf
28. Siani A, Tranter A (2022) Is vaccine confidence an unexpected victim of the COVID-19 pandemic? Vaccine 40(50):7262–7269
29. UNICEF (2023) The State of the World's Children: for every child, vaccination
30. Solis Arce JS, Warren SS, Meriggi NF, Scacco A, McMurry N, Voors M et al (2021) COVID-19 vaccine acceptance and hesitancy in low- and middle-income countries. Nat Med 27(8):1385–1394
31. Patwary MM, Alam MA, Bardhan M, Disha AS, Haque MZ, Billah SM (2022) COVID-19 vaccine acceptance among low- and lower-middle-income countries: a rapid systematic review and meta-analysis. Vaccines 10(3):427
32. Bergen N, Kirkby K, Fuertes CV, Schlotheuber A, Mac Feely S, Hosseinpoor AR et al (2023) Global state of education-related inequality in COVID-19 vaccine coverage, structural barriers, vaccine hesitancy, and vaccine refusal: findings from the global COVID-19 trends and impact survey. Lancet Glob Health 11(2):e207–e217
33. Siani A, Carter I, Moulton F (2022) Political views and science literacy as indicators of vaccine confidence and COVID-19 concern. J Prev Med Hyg 63(2)
34. Kibongani Volet A, Scavone C, Catalán-Matamoros D, Capuano A (2022) Vaccine hesitancy among religious groups: reasons underlying this phenomenon and communication strategies to rebuild trust. Front Public Health 10
35. Larson HJ, Schulz WS, de Figueiredo A, Jones NS, Xiahong Z, Cook AR et al (2016) The state of vaccine confidence 2016: global insights through a 67-country survey. EBioMedicine 12:295–301
36. Jones DG (2022) Religious concerns about COVID-19 vaccines: from abortion to religious freedom. J Relig Health 61(3):2233–2252
37. Boro E, Sapra T, de Lavison J-F, Dalabona C, Ariyaratne V, Samsudin A (2022) The role and impact of faith-based organisations in the management of and response to COVID-19 in low-resource settings. Relig Develop (2750–7947) 1(1):132–146
38. Siani A, Green I (2023) Scientific misinformation and mistrust of COVID-19 preventive measures among the UK population: a pilot study. Vaccines 11(2)
39. Zimmerman T, Shiroma K, Fleischmann KR, Xie B, Verma N, Lee MK et al (2023) Misinformation and COVID-19 vaccine hesitancy. Vaccine 41(1):136–144
40. Lee SK, Sun J, Jang S, Connelly S (2022) Misinformation of COVID-19 vaccines and vaccine hesitancy. Sci Rep 12(1):1–11
41. Cha M, Lima G, Singh K, Cha C, Ahn YY, Kulshrestha J et al (2021) Prevalence of misinformation and factchecks on the COVID-19 pandemic in 35 countries: observational infodemiology study. JMIR Hum Factors 8(1)

42. Dash S, Parray AA, De Freitas L, Mithu MIH, Rahman MM, Ramasamy A et al (2021) Combating the COVID-19 infodemic: a three-level approach for low and middle-income countries. BMJ Global Health 6(1)
43. Moola S, Gudi N, Nambiar D, Dumka N, Ahmed T, Sonawane IR et al (2021) A rapid review of evidence on the determinants of and strategies for COVID-19 vaccine acceptance in low- and middle-income countries. J Global Health 11
44. World Health Organization (2020) Global partnership to make available 120 million affordable, quality COVID-19 rapid tests for low- and middle-income countries. Available from: https://www.who.int/news/item/28-09-2020-global-partnership-to-make-available-120-million-affordable-quality-covid-19-rapid-tests-for-low--and-middle-income-countries
45. ACTaccelerator (2020) The access to COVID-19 tools accelerator. Available from: https://www.act-a.org/about
46. Storeng KT, de Bengy PA, Stein F (2023) COVAX and the rise of the 'super public private partnership' for global health. Glob Public Health 18(1):1–17
47. Pushkaran A, Chattu VK, Narayanan P (2023) A critical analysis of COVAX alliance and corresponding global health governance and policy issues: a scoping review. BMJ Global Health 8(10)
48. Das JK, Lakhani S, Khan MH, Bhutta ZA, Muhammad S, Chee HY et al (2023) COVID-19 vaccines: how efficient and equitable was the initial vaccination process? Vaccines 11(1)
49. Levine AC, Park A, Adhikari A, Alejandria MCP, Bradlow BH, Lopez-Portillo MF et al (2023) The role of civil society organizations (CSOs) in the COVID-19 response across the global south: a multinational, qualitative study. PLOS global public health 3(9):e0002341
50. World Health Organization (2023) COVAX—Working for global equitable access to COVID-19 vaccines. Available from: https://www.who.int/initiatives/act-accelerator/covax
51. Brown S, Rosier M (2023) COVID-19 vaccine apartheid and the failure of global cooperation. Br J Polit Int Relat 25(3):535–554
52. Lanziotti VS, Bulut Y, Buonsenso D, Gonzalez-Dambrauskas S (2022) Vaccine apartheid: this is not the way to end the pandemic. J Paediatr Child Health 58(2):228–231
53. Moore S, Hill EM, Dyson L, Tildesley MJ, Keeling MJ (2022) Retrospectively modeling the effects of increased global vaccine sharing on the COVID-19 pandemic. Nat Med 28(11):2416–2423
54. Ledford H (2022) COVID vaccine hoarding might have cost more than a million lives. Nature. https://doi.org/10.1038/d41586-022-03529-3
55. Gleeson D, Tenni BF, Townsend B, Phillips T (2023) Global inequities in access to COVID-19 health products and technologies: a political economy analysis. Health Place 83

7 Isoperimetric Control Inequality: Application to the Management of Limited Resources in Times of Pandemics

Ilias Elmouki, Albatoul Khaled, Amine Hamdache, Ling Zhong, Abdelilah Jraifi, and Aziz Darouichi

"An ounce of prevention is worth a pound of cure."
Benjamin Franklin.

Abstract

According to numerous past epidemic outbreaks, a poorly controlled pandemic can have catastrophic effects, killing millions of people in a matter of decades and having a significant effect on demography. When an unprecedented epidemic suddenly goes to the global level, it becomes more difficult to predict how the existing health and economic resources will succeed in managing the situation especially when the number of deceased people increases quasi-exponentially as happened during the COVID-19 time. In this chapter, we try to answer the question: how is it possible to devise the best metapopulation control management plans that take into account the many health resource constraints that decision-makers may encounter during pandemic emergencies? For that, we state in this chapter, the fundamental rules of optimal control strategies that are subject to multi-group-like models and which for the particularity

I. Elmouki (✉)
Bureau de M. A. Jraifi, ENSA-S, Safi, Morocco
e-mail: i.elmouki@gmail.com

A. Khaled · A. Hamdache · A. Jraifi
Laboratory of Mathematics, Computer Science and Communication Systems (MISCOM), National School of Applied Sciences-Safi (ENSA-S), Safi, Morocco

L. Zhong
Department of Economics, Cheung Kong Graduate School of Business, Beijing, China
e-mail: lingzhong@ckgsb.edu.cn

A. Darouichi
Department of Computer Science, FST, Cadi Ayyad University, Marrakesh, Morocco

B. W. K. Son (ed.), *The Landscape of Global Health Inequity*, Integrated Science 22,
https://doi.org/10.1007/978-3-031-60502-4_7

of the topic, need in addition to relying on the frameworks of the recently devised multi-region control systems, the implementation of what we refer to isoperimetric control inequalities to make practical policies with the best administration of scarce resources. We provide different theoretical scenarios of this approach and a review analysis when it can be used in concrete examples as when considering dengue epidemic breakouts for a region within itself under budgetary restrictions on insecticide supply, or when addressing immunization-related shortages in two interconnected regions. In the end, we conclude that any control approach would be ineffective if the problems of the health system limitations and the lack of medical resources are not addressed.

Keywords

Pandemic • Epidemic • Health resource • Multi-region modeling • Optimal Control • Isoperimetric constraint • Dengue • COVID-19

1 Introduction

1.1 General Problem and Main Goal

Etymologically, the term pandemic derives from pan and demos, all and people, and Greeks used it to refer to diseases that affect the entire population. A poorly managed pandemic can have disastrous effects, as seen by the plague of 1347 [1], which killed millions of people in a few decades and significantly altered the population's demographics. In such scenarios, limited medical resources are also known to have direct and strong impacts on the transmission and spread of pandemics as was for instance, in the more recent case of the Yemen cholera outbreak between 2016 and 2022 [2].

Nowadays, the world is still facing the threat of seasonal coronaviruses and their variants. When there is a lack of information on the origin and nature of the pandemic, some would start to think that, in the early phases of clinical research, the only option would be to use advanced statistics and mathematics to at least comprehend the disease's spatiotemporal dynamics or take the initiative to anticipate it as was attempted with COVID-19 [3–7].

Modeling frameworks that have been used to examine several topics in ecology, immunology, and epidemiology utilizing compartmental systems as covered in the books [8, 9], frequently aided at the microscopic level in the analysis of how the infection moved from one host to another or, more broadly, from one cell to another. However, when it is about a pandemic, it makes more sense that the only models that could be used to analyze the dynamics of global epidemics are those that resemble metapopulation-like systems [10–12]. Even though these models are more useful for understanding pandemic conditions at a wide geographic scale, there is still a disinclination towards them. After such a model is validated, optimal control theory can be applied for the optimization of the anti-epidemic interventions, as was the case with the metadata models in [13–15]. It is also challenging to forecast how the current medical and financial resources would be able to handle an unprecedented epidemic that quickly spreads around the world, particularly when the death toll rises almost exponentially as it did during the COVID-19 pandemic, and to manage disparities to suggest workable strategies.

In this chapter, we attempt to address the following query: how can the most effective control management policies be developed while taking into account the different limitations that could result in a shortage of medical resources during pandemic crises? For that, we lay out in this chapter, the foundational principles of optimal control strategies that apply to our past multi-region pandemic systems, and because of the specificity of the subject, we explain how these principles also require the application of what we refer to as isoperimetric control inequalities to provide actionable policy recommendations. The methods in this chapter are interesting to advance the work completed in [8–21] and ignored the presence of such resource inequalities or constraints. In addition, we provide two theoretical examples: the first solves the continuous-time optimal control problem that aims to determine the optimal duration of the control policy subject to an epidemic model with a short-term immunized population while defining an isoperimetric limit due to a lack of adequate health resources, while the second solves a control problem subject to similar isoperimetric constraint but in a more suitable modeling framework devised for the discrete-time case.

We also evaluate our approach through the review of two additional concrete cases to help make it easier for academics to profit from it: the first deals with dengue epidemic outbreaks inside a specific region while adhering to budgetary constraints on insecticide supply, and the second handles vaccination-related shortages in two connected districts.

Finally, we conclude that any control strategy would be useless if the shortage of medical resources and the limits on the capacity of the health system were ignored. This is why developing new control plans with isoperimetric inequalities is essential. In the absence of that, some research would be restricted to classical statistical techniques, which are, as far as the authors are aware, not advanced enough or ready to handle the problems of medical and financial resources in the current era of networked big data, notwithstanding the application of novel learning methodologies.

1.2 General Context of Control for Epidemics

Affected regions were forced to adopt the most straightforward control measures that the majority of people saw as the rational interventions to succeed immediately in the war against an epidemic, without resorting to any complex method that would require time, a whole scientific theory, and sufficient evidence to support it [22] during most times of unknown pandemics whose causes are frequently unknown [23]. Reviving the notion that cleanliness is always necessary to achieve hygiene (personal cleanliness) and sanitation (public cleanliness) [24] is one of the control approaches that were advised to be respected when discussing the majority of cases of infectious diseases worldwide [25]. When an organism shows signs of being able to spread from one host to another, people will instinctively isolate infectious pathogens that exhibit clear symptoms. This has been a very old practice for centuries, and it was even used in circumstances that are not currently particularly harmful [26]. However, as the idea of population immunization has evolved, its usefulness has diminished [27]. Certain customs, like using vinegar as a disinfectant, were influenced by the wisdom of some ancient savants [28]. The advantages of alternative tactics, including isolating sick individuals or groups, were typically inferred from a body of information on historical epidemic episodes, prior experiences, and the economic conditions of various geographic areas [29, 30]. But in our new world, it cannot be that easy. There is no denying of the big problems caused by the world population's exponential increase. In actuality, the more complicated this environment gets, the harder it will be to find answers, particularly in the event of an epidemic that might spread quickly across a wide geographic area and appear out of the blue.

The growing number of intra-regional connections within a region due to population movements and the connections with other regions through air travel, for example, forced the scientific community to search for the best strategy that must consider numerous variables and constraints of different kinds and locations to prevent more unpredictable outcomes, in addition to looking for potentially traditional anti-epidemic measures that would reduce infection quickly. When applied to epidemic models and networks, these are some of the primary goals of novel optimal control approaches. It can also be demonstrated here that there is still a deficiency in the research of modeling and control science, which hinders its advancement and viability, particularly when it comes to a worldwide phenomenon.

Along with introducing and providing suitable controls and technical measures, this work also intends to complete our analysis of the practical applicability of the methods that we have discussed in [31, 32] and provide alternative strategies to gain a better understanding of the dynamics of emerging epidemics in the presence of the limited resource constraints.

1.3 Specific Developments of Control for Pandemics

The primary concept underlying the significance of optimal control in epidemiology is that it aids in proposing the most effective course of action to combat an infection subject to a specific dynamical system, which is utilized to characterize the spread of an epidemic from one state to another, as indicated by the mathematical Eqs. (1–5) in the following section utilizing the X vector, and/or moving across regions as indicated by the domain and subdomains in the same relations by Omega(s).

To elucidate this general idea, consider the following: the primary objective, which is typically to reduce the number of infectives while increasing the number of recovered individuals, which are components of X, can be expressed as an optimization problem, as mentioned in (2–5), with a predetermined objective function J of the state variables and the control function employing vector u. For example, consider the case of HIV/AIDS infection and acquired immune deficiency syndrome (AIDS) controlled by awareness [18]. Equations of discrete variations are used to describe the dynamics of these states, which could be the numbers of individuals or the proportions of the classes taken into consideration in a population [30] as in Eq. (1) for times $i = 0,\ldots,N\text{-}1$ or continuous variations [33] concerning time and/or space as in Eq. (4) for time t in [0, T] and regions $j = 1,\ldots,p$, and that serve as the primary restriction on this optimization as stated in (2–3, 5). The optimal control issue gets harder to solve as dynamics become more complex, as shown in (4–5). After identifying the best control values, the control function applied to these dynamics concretely depicts the intervention selected to shift susceptibility, infectivity, and/or recovery to the best controllable states.

Since the authors of [34] were among the first to publish precise algorithms in this regard, we often advise scholars interested in devising optimal control policies for more complex theoretical scenarios to start from that reference for clear mathematical and numerical details. Before delving into more detail about this theory, it should be noted that, in the case of COVID-19, many researchers frequently referred to the control non-pharmaceutical interventions (NPIs) [35–55], which demonstrated their effectiveness in parallel pandemics like influenza [50], but if implemented sooner for better economic and health outcomes [52]. This was due to the lack of a treatment or vaccine. As previously mentioned, the research approach we are working with here necessitates a precise modeling framework where the control could be implemented. Davies et al. [53] employed an age-structured transmission model, always in the context of NPIs against the spread of COVID-19, and discovered that such interventions are crucial steps in lowering the number of deaths and demands on hospital beds.

Regarding the best control policy at this time, Zamir et al. in [54] determined the best COVID-19 policy and concluded that NPIs should be maintained; otherwise, fresh outbreaks will inevitably recur. Ullah et al. [55] modeled the kinetics of COVID-19 in a different study. An optimal control analysis was employed to examine the effects of NPIs. They suggested focusing on social separation and quarantine to prevent the worst-case scenario of the illness spreading based on their findings. In her work, Jasmina Panovska-Griffiths concluded [56] that new models need to be created by researchers because it does not seem possible to gain all the answers using only one or a small number of the existing frameworks, particularly when discussing the perplexing circumstances brought on by the COVID-19 pandemic.

We believe that in most epidemic differential systems, the two infrequent factors that are still not sufficiently studied and mathematically detailed are the modeling of individual movement and the modeling of inevitable perturbations, such as rumors, which could thwart most strategies. As we will discuss in the section that follows, some recent advancements in modeling and optimum control techniques have been offered in this context.

2 Contributions to Global Epidemic Control Modeling

The effect of human motion on COVID-19 was covered in many articles. Kraemer et al., for example, demonstrated [57] the value of travel restrictions during the initial phases of the outbreak in China. Carteni et al. [58] provided a statistical explanation of the direct relationship between human mobility habits and the number of COVID-19 infections in Italy through the use of a multiple linear regression model. Zhu et al. [59] discovered in their investigation of the spatial–temporal potential exposure risk of COVID-19 that movement constraints enhance air quality by lowering industrial production, transportation, and traffic.

Recently, there has been an increased interest among researchers to simulate the spatial spread of pandemics utilizing basic equations derived from classical epidemic models, such as the discrete [60–62] or continuous [63, 64] models. However, since numerous connections made it easier for the disease to spread among the three affected countries, we can refer to the multi-region continuous-time models applied to the Ebola case in [19] for the description of travel in a wide geographic area, taking into account new factors like the movement of infectious agents from one region to another. HIV/AIDS [20] was a comparable situation, as noted by intra-connectivity within a region due to people migration within cities, villages, and neighborhoods. A similar concept was previously employed by Arino and Van Den Driessche [64] in this continuous-time scenario as in [20], but they took into account the mobility of all classes, including the susceptible, exposed, and recovered population. Afterward, with the advent of a discrete optimal control approach, Zakary et al. [65] noted that especially when attempting to target and control a region, the class that needs to be controlled the fastest and is more accountable for a pandemic's spread from one domain to another is the infective population alone, both through their movements and, of course, through their interactions with other categories. It is preferable to use discrete-time models over continuous-time ones. The theoretical foundation for these discoveries was built upon more straightforward systems in the form of discrete difference equations that were initially developed in [66] and which were discrete because, as is further indicated in [30], the writers honored the notion that data were often acquired at discrete times.

To sum up, the first discrete-time multi-region epidemic models' general form [65, 66] is

$$X_{i+1}^{\Omega} = X_i^{\Omega} + f\left(X_i^{\Omega}, u_i^{\Omega}\right), i = 0, \ldots, N-1 \tag{1}$$

The primary limitation of the discrete optimal control problem is as follows:

$$J\left(u_j^{\Omega_j}\right) = \Phi\left(N, X_N^{\Omega_j}\right) + \sum_{i=0}^{N-1} f_0\left(X_i^{\Omega_j}, u_i^{\Omega_j}\right) \tag{2}$$

Likewise, by substituting the integral for the sum and the derivative for the variation with respect to time, the generic form of the continuous optimal control problem is

$$J\left(u_j^{\Omega_j}\right) = \Phi\left(T, X^{\Omega_j}(T)\right) + \int_0^T f_0\left(t, X^{\Omega_j}(t), u^{\Omega_j}(t)\right)dt \tag{3}$$

with respect to types of multi-region, continuous-time epidemic models devised in [19, 20].

Later research benefited from the work done in references [65, 66] and applied their modeling and control techniques to other systems [12–14, 64–70] using a new type of cellular simulations for epidemics. This demonstrated the effectiveness of the optimal control, which gave the best proportions of the mobility restriction of infected travelers only, as the authors assumed no other intervention due to the unavailability of any medical treatment or vaccination. It also highlighted the importance of infection mobility in the spatial propagation of an epidemic in a large domain.

In addition to travel, rumors are a significant factor in the current spatial spread of infection and are frequently disregarded. Rumors can impede the effectiveness of any control measures taken by the relevant authorities. Most individuals now find it simpler to disseminate rumors that can spread quicker at the same time as an epidemic thanks to modern technology and the existence of online social networks. Infodemics are these rumors spreading like wildfire. To the best of our knowledge, doctor Briand, director of Infectious Hazards Management at the World Health Organization (WHO)'s Health Emergencies Programme and architect of WHO's strategy, was the first scientist to use this term more frequently in the scientific community and to issue a warning about it. A similar opinion was reported in [71]. Scientists therefore became more conscious of the infodemic problem as COVID-19 spread and worked to develop countermeasures [72–78].

In a joint statement [111] issued on September 23, 2020, the WHO, United Nations (UN), United Nations Children's Fund (UNICEF), United Nations Development Programme (UNDP), United Nations Educational, Scientific and Cultural Organization (UNESCO), United Nations Programme on HIV and AIDS (UNAIDS), International Telecommunication Union (ITU), UN Global Pulse, and International Federation of Red Cross and Red Crescent Societies (IFRC) advised countering misinformation and disinformation to contain the COVID-19 epidemic. Indeed, even within the framework of our main topic, which addresses the issue of managing scarce resources, an infodemic may cause a frantic abuse of medical services and render hospitals incapable of covering the escalating demands of treatments.

As noted in theoretical contexts in [19] and empirical contexts in [83, 84], mistrust may give rise to misunderstandings and concerns in the context of the Ebola outbreak in Sierra Leone, and certain beliefs may be based on unsubstantiated scientific claims. Eliza Y.L. Cheung, an IFRC psychosocial delegate dispatched to Liberia amid the Ebola crisis said in [85] that before offering health advice or assistance, it is critical to understand the culture and perceptions of the affected population and that over-protection could make matters worse.

In addition to the deterministic conclusions in [13], El Kihal et al. in [12] and [86] were the first to create new stochastic control models for the study of infodemic effect and a broad mathematical framework in the topic of the geographical spread of pandemics under perturbations.

$$dX^{\Omega}(t) = f\left(t, X^{\Omega}(t), u^{\Omega}(t)\right)dt + g\left(t, X^{\Omega}(t), u^{\Omega}(t)\right)dW^{\Omega}(t), t \in [0, T] \quad (4)$$

and the related form of the objective function in the stochastic continuous-time scenario is displayed as

$$J\left(u_j^{\Omega_j}\right) = \mathbb{E}\left(\Phi\left(T, X^{\Omega_j}(T)\right) + \int_0^T f_0\left(t, X^{\Omega_j}(t), u^{\Omega_j}(t)\right)dt\right) \quad (5)$$

If the spatial aspect of pandemics discussed in [65, 66] and the use of optimal control theory to ascertain the optimal values of the anti-epidemic measures put in place are not taken into account, stochasticity could provide a strong theoretical basis for the mathematical modeling of global epidemics. For instance, the authors in [87] investigated how the environment affects COVID-19 transmission and provided more realistic research by developing a stochastic epidemic model of the virus. In the same context, authors in [88] studied the impacts of environmental factors, such as humidity, precipitation, and temperature, on the spread of a new strain of coronavirus COVID-19 to humans through a different stochastic mathematical model.

3 Isoperimetric Inequality in the Administration of Scarce Resources

3.1 Introduction with Limited Vaccine Stock

Along with all the optimal control strategies that have been covered thus far, we also recommend in this chapter a few developments that, in our opinion, are more realistic as they should be considered regardless of how difficult they may turn out when they are subject to the control systems and objective functions shown in the aforementioned equations while introducing isoperimetric inequalities that would help to find better control plans that often meet the constraints on financial and health resources. To make it simpler for researchers who would be interested in investigating our approach that is recommended to apply to the above equations, we deal with less difficult cases to understand how such isoperimetric inequalities could be defined. We can start by defining the case where vaccine stocks are limited and find the optimal vaccination schedule such that only a given number of people are vaccinated. For example, take a Susceptible-Exposed-Infected-Removed (SEIR) model as in [34] to illustrate the use of optimal control.

The exposed class includes those who have the disease but remain latent for some time before becoming contagious. The total population at time t is represented by

$$N(t) = S(t) + E(t) + I(t) + R(t).$$

The differential equations governing these classes are,

$$\begin{aligned} S'(t) &= bN(t) - dS(t) - cS(t)I(t) - u(t)S(t) \\ E'(t) &= cS(t)I(t) - (e + d)E(t) \\ I'(t) &= eE(t) - (g + a + d)I(t) \\ R'(t) &= gI(t) - dR(t) + u(t)S(t) \\ N'(t) &= (b - d)N(t) - aI(t) \end{aligned}$$

where the control denoted by u(t) represents the percentage of susceptible individuals vaccinated per unit of time. It is assumed that the vaccine is always effective, resulting in all vaccinated individuals becoming immune. The control for example is limited between two bounds to prevent the scenario where the entire susceptible population is vaccinated. Let b be the natural exponential birth rate of the population and d be the natural exponential death rate. The disease incidence or frequency is described by the term cS(t)I(t). The parameter e represents the rate at which exposed individuals become infectious, and g is the rate at which infectious individuals recover. The parameter a denotes the mortality rate due to the disease in an infected individual. To apply optimal control to a model for a specific disease, it must be verified for the uncontrolled model that the basic reproductive number R_0 is greater than one for the parameter ranges considered. Therefore, disease eradication without intervention strategies is not expected.

Now, and this is what interests us more here, is that one could extend this model to consider a scenario where a limited vaccine supply is available. If X represents the total number of susceptible individuals we can vaccinate over T years, then we have an integral constraint,

$$\int^{T} S(t)u(t)dt \le X$$

These constraints are known as isoperimetric constraints and can be handled by adding another state variable, W(t), such that

$$\begin{aligned} W'(t) &= S(t)u(t) \\ W(0) &= 0 \\ W(T) &\le X \end{aligned}$$

The main goal will be to minimize the number of infected individuals and the overall cost of vaccination over T years.

As in [34], the problem may be defined as,

$$\int_{0}^{T} AI + (t)u^2(t)dt$$

With A the constant severity weight and with the conditions,

$$S(0) = S_0, E(0) = E_0, \quad I(0) = I_0, \; N(0) = N_0$$

And,

$$S(T), E(T), I(T), \; N(T) \text{ are fixed.}$$

3.2 Budgetary Restrictions on Insecticide Supply

Control intervention practices usually carried out by health authorities in tropical areas suffer from occasional dengue outbreaks in the city of Cali, Colombia. Such control interventions as reported in [89], are mainly based on insecticide spraying, and sporadically conducted to overcome an ongoing epidemic or at least reduce its magnitude. These interventions are crucial due to the risks associated with this disease. In fact, the risk of contracting dengue varies seasonally and annually, with mosquitoes typically biting during the day, especially around sunrise and sunset. Since there is no specific preventive treatment for dengue, it is essential to take precautions against mosquito bites.

The authors in [89] examined three scenarios (20, 40, and 60%) of budget reduction related to the supply of insecticides in combination with three types of insecticides with low (Tebufenozide), medium (Permethrin), and high lethality (Organochlorines or Carbamates). Then, each scenario is denoted as Scenario i-j with i indicating the strategy i = 1, 2, 3 corresponding to the insecticides chosen in each lethality category. and j the case of resource limitation j = 1, 2, 3, and this requires to solve the following optimal control problem,

$$\int_{0}^{T} C_1 H(t) + \frac{C_2}{2} u^2(t)dt$$

subject to

$$\frac{dV}{dt} = ap_V H(1 - V) - [b + u(t)]V, V_0 = V(0)$$
$$\frac{dH}{dt} = ap_H dV(1 - H) - [b + u(t)]V, H_0 = H(0)$$

$$\frac{dC_h}{dt} = ap_H dV(1-H) - cH, H_0 = C_h(0)$$
$$\frac{dZ}{dt} = C_2 u(t), Z(0) = 0, Z(T) = B$$

where $C_1 > 0$ is the average daily societal cost of having one infected human individual (eg, treatment, temporary disability, etc.) and let $C_2 > 0$ be the average daily cost of control intervention $u(t)$ (eg, insecticide and spraying supplies, manpower costs, etc.), and where V(t) is the fraction of infected mosquitos at the moment t, H(t) is the fraction of infected human hosts at the moment t, u(t) is the mosquito mortality rate due to insecticide spraying at the moment t, average number of female mosquitoes per one human host, a is the human blood feeding rate, the proportion of mosquitoes that feed on humans each day, b is the rate of mosquito natural mortality, c is the rate of human recovery from the disease, d is the average number of female mosquitoes per one human host, p_V is the proportion of infected human hosts that are infectious, p_H is the proportion of infected mosquitoes that are infectious.

Then, after defining the objective and its subject, there is the implementation of isoperimetric constraint which serves to define the limited amount of external resources B, and that takes the form of the following inequality,

$$\int_0^T C_2(t)u(t)dt \leq B$$

Specifically, the additional benefits brought by the optimal control policy obtained are insignificant when the low-lethality insecticide is applied either without budget constraints or within budget reductions of 20–40%. This implies that the local healthcare system can save between \$9,600 and \$101,400 by solely changing its insecticide spraying operational mode. In a low-income country like Colombia, where over 20% of children under 5 years lack access to basic vaccination plans, this is an important consideration.

3.3 Immunization-Related Shortages in Two Interconnected Regions

In a resource-constrained environment, the authors in [87] propose and study an epidemic model consisting of two host groups to reflect varying levels of infection risk and transmission.

$$\frac{dS^r}{dt} = A^r - \left(B_r^r \frac{I^r}{1 + a^r I^r} + B_c^r \frac{I^r}{1 + a^c I^c}\right) S^r - \left(b + c^r u_1\right) S^r$$
$$\frac{dS^c}{dt} = A^c - \left(B_r^c \frac{I^r}{1 + a^r I^r} + B_c^c \frac{I^r}{1 + a^c I^c}\right) S^c - \left(b + c^c u_2\right) S^c$$

$$\frac{dI^r}{dt} = \left(B_r^r \frac{I^r}{1 + a^r I^r} + B_c^r \frac{I^r}{1 + a^c I^c}\right) S^r$$
$$+ \left(K_r^r \frac{I^r}{1 + a^r I^r} + K_c^r \frac{I^r}{1 + a^c I^c}\right) V^r - (d + v^r) I^r$$
$$\frac{dI^c}{dt} = \left(B_c^c \frac{I^r}{1 + a^r I^r} + B_r^c \frac{I^r}{1 + a^c I^c}\right) S^r$$
$$+ \left(K_r^c \frac{I^r}{1 + a^r I^r} + K_c^c \frac{I^r}{1 + a^c I^c}\right) V^c - (d + v^c) I^c$$
$$\frac{dV^r}{dt} = c^r u_1 S^r \left(K_r^r \frac{I^r}{1 + a^r I^r} + K_c^r \frac{I^r}{1 + a^c I^c}\right) V^r - dv^r$$
$$\frac{dV^c}{dt} = c^c u_2 S^c \left(K_r^c \frac{I^r}{1 + a^r I^r} + K_c^c \frac{I^r}{1 + a^c I^c}\right) V^c - dv^c$$

where $S^r(t)$ is the number of risky susceptible individuals in t time, S^c (t) is the number of critical susceptible individuals in t time, I^r (t) is the number of risky infected individuals in t time, I^C (t) is the number of critical infected individuals in t time, V^r (t) is the is the number of risky vaccinated individuals in t time, V^C (t) number of critical vaccinated individuals in t time, B_j^i is the contact rate between susceptible individuals and infected individuals (i, j = r, c), K_j^i is the contact rate between vaccinated individuals and infected individuals (i, j = r, c), v^i is the disease-induced death rate (i = r, c), d is the natural death rate, c^i is the rate at which susceptible individuals become successfully vaccinated (i = r, c)

In the previous system, u_1 and u_2 respectively denote the coverage rates of individuals S^r and S^c who are vaccinated. t_f is the final time, and the control functions u_i (t), i = 1, 2 are defined on the closed time interval [0, t_f]. The control functions u_1, u_2 are bounded and Lebesgue integrable. The coefficients C_1, C_2 separately denote the corresponding weight constants that balance the cost elements based on their size and importance in the parts of the objective function.

The main goal of the study is to reduce the number of infectious individuals and the cost of implementing the strategy.

$$\int_0^t C_1 I^r + C_2 I^c + e_1 u_1^2 + e_2 u_2^2 dt$$

In cases of limited control resources with constant and variable coverage rate strategies, it is demonstrated that when total control resources are scarce, resources should be directed towards the group with a high infection risk level; and control resources should preferably be directed towards the low-risk group if total resources are relatively abundant.

Now, we provide a simple numerical example of the isoperimetric control approach when applied to an epidemic with short-term immunity, based on the optimal control approach but with the slight difference in this chapter that we are

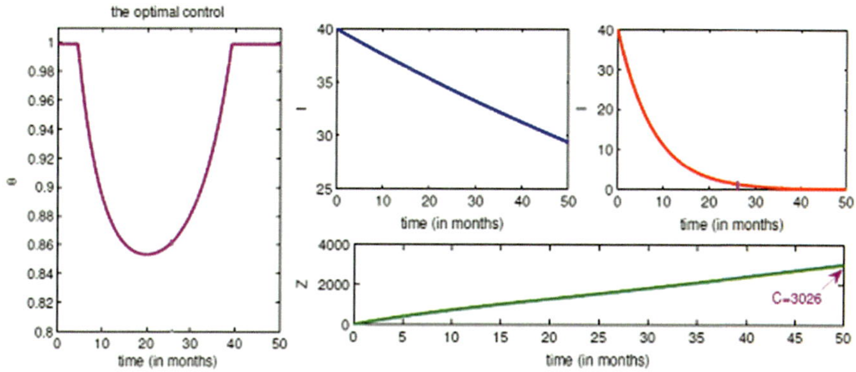

Fig. 1 The optimal control

not treating the problem of the optimal control policy duration, then by considering the subject of the objective as a differential SIR epidemiological system with a control θ introduced as vaccination, and with the definition of the additional isoperimetric constraint,

$$\int_0^T \Theta(t)u(t)dt \leq C$$

meaning that we can only vaccinate a number not more than for example C = 3026, thus, we find from Fig. 1 that we can reach our objective by reducing the number of infected people represented by the I function while the new variable Z represents the limitation on the vaccine stock, gets a solution that could not extend the constant C.

Note. Purple: The optimal control in the presence of isoperimetric inequality. Blue: The Infective function I without any control and/or constraint. Red: The Infective function I in the presence of the isoperimetric control inequality. Green: The associated variable Z to the isoperimetric inequality reaching the imposed constant C.

4 Discussions

When it comes to unprecedented pandemics, it is not really that convincing to ignore that many locations suffer from many limitations in their resources. Take the example of COVID-19 crisis, there was a decrease in the human resources available to care for a population. Key messages also in [91] were about supplemental oxygen as a first essential step for treating severe COVID-19 patients with hypoxemia as the capacity for invasive ventilation was limited. Other observations in densely populated countries like Bangladesh [92] were about the weak governance in the fragile healthcare system as there were for example no dedicated hospital units for COVID-19 patients.

Concerned that demand was surpassing capacity during COVID-19 in the United States, elected authorities, medical professionals, and healthcare organizations called at that time for more funding and resources [93]. There may be different reasons that could lead to such situations, as in Italy, the paucity of resources brought on by budget cuts, results in an inadequate number of spots in intensive care units [94]. When medical professionals are not prepared for global epidemics, they may be compelled to select which patients to treat due to a lack of resources [95].

Nowadays, there are a variety of optimal control approaches that can be used to stop the spread of an infectious disease. It is now possible to determine the best health strategy in times of epidemic emergencies while relying on different frameworks such as deterministic control systems in [96–103], stochastic control models as in [12, 86], and [104], or with more recently developed impulse control principles as in [105–107]. However, to limit the effects of limitations in health resources, future perspectives should focus on the consideration of isoperimetric inequalities as in [30] and [108–110] and as developed in this chapter.

Even though these constraints may lead to issues that are less mathematically and numerically tractable, it is still important to define such control constraints and not overlook them, as concretely and according to [111], control strategies that reduce demand for medical supplies, increase supply and create new supplies, can help lessen the likelihood of significant challenges in the future, and this is particularly important for both epidemic and pandemic issues.

5 Conclusion

This chapter started with the presentation of the most recent optimal control designs to different mathematical models of pandemics, then, it showed the importance of isoperimetric inequalities through many theoretical and concrete situations and where there have been considerations of the limitations on health and budget resources. This is very important for both local and global epidemic issues as control policies that lower for example the demand for medical materials, boost supply, generate new supply, as well as the staff that supply appropriately and all the necessary budgetary resources to succeed in such management operations, can help lessen the likelihood of significant challenges in the future. In fact, as opposed to the case of their absence and which is the most adopted optimal control framework since most control devisers would find it more tractable mathematically and numerically, these inequalities have proven through the examples presented in this chapter, their potential to produce better control plans for future pandemic-related management issues.

Core Messages

- Any control strategy would be worthless if efforts were not made to address the capacity constraints of many health systems and the scarcity of medical resources, and if not thinking about serious solutions to the huge gaps that exist between different regions.

References

1. Christakos G, Olea RA, Serre ML, Wang LL, Yu HL (2005) In: Interdisciplinary public health reasoning and epidemic modelling: the case of black death. Springer. ISBN 978-3-540-25794-3
2. He D, Wang X, Gao D, Wang J (2018) Modeling the 2016–2017 Yemen cholera outbreak with the impact of limited medical resources. J Theor Biol 451:80–85
3. Arenas A, Cota W, Gómez-Gardeñes J, Gómez S, Granell C, Matamalas JT, Soriano-Paños D, Steinegger B (2020) Modeling the spatiotemporal epidemic spreading of COVID-19 and the impact of mobility and social distancing interventions. Phys Rev X 10(4):041055
4. Viguerie A, Lorenzo G, Auricchio F, Baroli D, Hughes TJ, Patton A, Reali A, Yankeelov TE, Veneziani A (2021) Simulating the spread of COVID-19 via a spatially-resolved susceptible-exposed-infected-recovered-deceased (SEIRD) model with heterogeneous diffusion. Appl Math Lett 111:106617
5. Wang Z, Zhang X, Teichert GH, Carrasco-Teja M, Garikipati K (2020) System inference for the spatio-temporal evolution of infectious diseases: Michigan in the time of COVID-19. Comput Mech 66(5):1153–1176
6. Zhou Y, Wang L, Zhang L, Shi L, Yang K, He J, Zhao B, Overton W, Purkayastha S, Song P (2020) A spatiotemporal epidemiological prediction model to inform county-level COVID-19 risk in the United States. Special Issue 1-COVID-19: Unprecedented Challenges and Chances
7. Brauer F, Castillo-Chavez C, Feng Z (2019) Mathematical models in epidemiology, vol 32. Springer, New York
8. Brauer F, Castillo-Chavez C (2012) In: Mathematical models in population biology and epidemiology. vol 2. New York, Springer, pp 508
9. Bichara D, Iggidr A (2018) Multi-patch and multi-group epidemic models: a new framework. J Math Biol 77(1):107–134
10. Arino J (2009) Diseases in metapopulations. In: Modeling and dynamics of infectious diseases. pp 64–122
11. Arino J, Van den Driessche P (2006) Disease spread in metapopulations. Fields Institute Communications vol 48(1). pp 1–13
12. El Kihal F, Abouelkheir I, Rachik M, Elmouki I (2019) Role of media and effects of infodemics and escapes in the spatial spread of epidemics: a stochastic multi-region model with optimal control approach. Mathematics 7(3):304
13. El Kihal F, Rachik M, Zakary O, Elmouki I (2017) A multi-regions SEIRS discrete epidemic model with a travel-blocking vicinity optimal control approach on cells. Int J Adv Appl Math Mech 4(3):60–71
14. Zakary O, Rachik M, Elmouki I (2017) A new epidemic modeling approach: Multi-regions discrete-time model with travel-blocking vicinity optimal control strategy. Infectious Disease Modelling 2(3):304–322
15. Peyravi M, Marzaleh MA, Shamspour N, Soltani A (2020) Public education and electronic awareness of the new coronavirus (COVID-19): experiences from Iran. Disaster Med Public Health Preparedness 1

16. Alahdal H, Basingab F, Alotaibi R (2020) An analytical study on the awareness, attitude and practice during the COVID-19 pandemic in Riyadh, Saudi Arabia. J Infect Public Health 13(10):1446–1452
17. Samanta S, Rana S, Sharma A, Misra AK, Chattopadhyay J (2013) Effect of awareness programs by media on the epidemic outbreaks: a mathematical model. Appl Math Comput 219(12):6965–6977
18. Zakary O, Rachik M, Elmouki I (2016) On the impact of awareness programs in HIV/AIDS prevention: an SIR model with optimal control. Int J Comput Appl 133(9):1–6
19. Zakary O, Rachik M, Elmouki I (2017) A multi-regional epidemic model for controlling the spread of Ebola: awareness, treatment, and travel-blocking optimal control approaches. Mathemat Methods in the Appl Sci 40(4):1265–1279
20. Zakary O, Larrache A, Rachik M, Elmouki I (2016) Effect of awareness programs and travel-blocking operations in the control of HIV/AIDS outbreaks: a multi-domains SIR model. Adv Difference Equ 2016(1):169
21. Hays JN (2005) Epidemics and pandemics: their impacts on human history. Abc-clio
22. Speltini G, Passini S (2014) Cleanliness/dirtiness, purity/impurity as social and psychological issues. Cult Psychol 20(2):203–219
23. Rosenberg CE (1992) Explaining epidemics. Cambridge University Press
24. Davies ML, Davies TL (1987) Biblical ills and remedies. J R Soc Med 80(8):534
25. Conant J (2005) Sanitation and cleanliness for a healthy environment. Hesperian Foundation
26. Goodall EW (1934) A short history of the epidemic infectious diseases. John Bale, Sons and Danielsson, Limited
27. Orey C (2008) The healing powers of vinegar-revised and updated. Kensington Publishing Corp
28. Conti AA (2008) Quarantine through history. In: International encyclopedia of public health, pp 454
29. Gensini GF, Yacoub MH, Conti AA (2004) The concept of quarantine in history: from plague to SARS. J Infect 49(4):257–261
30. El Kihal F, Abouelkheir I, Rachik M, Elmouki I (2018) Optimal control and computational method for the resolution of isoperimetric problem in a discrete-time SIRS system. Mathem Computat Appl 23(4):52
31. Elmouki I, Zhong L, Jraifi A, Darouichi A (2023) Optimal control: application and applicability in times of pandemics. Integrated science of global epidemics. Springer International Publishing, Cham, pp 191–210
32. Rezaei N et al. (2023) Integrated science of global epidemics 2050. In: Rezaei (ed) Integrated science of global epidemics, Springer International Publishing, pp 587–607
33. Abouelkheir I, El Kihal F, Rachik M, Elmouki I (2018) Time needed to control an epidemic with restricted resources in SIR model with short-term controlled population: a fixed point method for a free isoperimetric optimal control problem. Mathemat Computat Appl 23(4):64
34. Lenhart S, Workman JT (2007) Optimal control applied to biological models. CRC Press
35. Shimul SN, Hussain M, Faisel AJ, Hamid SA, Sultana N, Kuddus MA (2024) Impact of alternative non-pharmaceutical interventions strategies for controlling COVID-19 outbreak in Bangladesh: a modeling study. PLoS ONE 19(2):e0293863
36. Imai N, Gaythorpe KA, Abbott S, Bhatia S, van Elsland S, Prem K, Liu Y, Ferguson NM (2020) Adoption and impact of non-pharmaceutical interventions for COVID-19. Wellcome Open Research 5(5):59
37. Flaxman S, Mishra S, Gandy A, Unwin HJT, Coupland H, Mellan TA, Zhu H, Berah T, Eaton JW, Guzman PN, Schmit N (2020) Report 13: estimating the number of infections and the impact of non-pharmaceutical interventions on COVID-19 in 11 European countries. Imperial College London. https://doi.org/10.25561/77731
38. Flaxman S, Mishra S, Gandy A, Unwin HJT, Mellan TA, Coupland H, Whittaker C, Zhu H, Berah T, Eaton JW, Monod M (2020) Estimating the effects of non-pharmaceutical interventions on COVID-19 in Europe. Nature 584(7820):257–261

39. Hens N, Vranck P, Molenberghs G (2020) The COVID-19 epidemic, its mortality, and the role of non-pharmaceutical interventions. Eur Heart J Acute Cardiovasc Care 9(3):204–208
40. Patiño-Lugo DF, Vélez M, Velásquez Salazar P, Vera-Giraldo CY, Vélez V, Marín IC, Ramírez PA, Quintero SP, Castrillón Martínez E, Pineda Higuita DA, Henandez G (2020) Non-pharmaceutical interventions for containment, mitigation and suppression of COVID-19 infection. Colombia Médica 51(2):e-4266
41. Lin X (2020) Analysis of 25,000 lab-confirmed COVID-19 cases in Wuhan: epidemiological characteristics and non-pharmaceutical intervention effects. Harvard University and Broad Institute, Department of Biostatistics and Department of Statistics
42. Ferguson N, Laydon D, Nedjati Gilani G, Imai N, Ainslie K, Baguelin M, Bhatia S, Boonyasiri A, Cucunuba Perez ZULMA, Cuomo-Dannenburg G, Dighe A (2020) Report 9: impact of non-pharmaceutical interventions (NPIs) to reduce COVID19 mortality and healthcare demand. Imperial College London 10:77482
43. Wong M, Huang J, Teoh J, Wong S (2020) Evaluation on different non-pharmaceutical interventions during COVID-19 pandemic: an analysis of 139 countries. J Infect 81(3):e70–e71
44. Seale H, Dyer CE, Abdi I, Rahman KM, Sun Y, Qureshi MO, Dowell-Day A, Sward J, Islam MS (2020) Improving the impact of non-pharmaceutical interventions during COVID-19: examining the factors that influence engagement and the impact on individuals. BMC Infect Dis 20(1):1–13
45. Haushofer J, Metcalf CJE (2020) Evaluation of non-pharmaceutical interventions is needed to mitigate the COVID-19 pandemic. Working Paper of Princeton University
46. Castex G, Dechter E, Lorca M (2020) COVID-19: Cross-country heterogeneity in effectiveness of non-pharmaceutical interventions. Covid Econom 14:175–199
47. KRR G, Casella F (2020) Non-pharmaceutical interventions (NPIs) to reduce COVID-19 mortality. Available at SSRN 3560688
48. Cho SWS (2020) Quantifying the impact of non-pharmaceutical interventions during the COVID-19 outbreak-the case of Sweden. Economet J 23(3):323–344
49. Al Wahaibi A, Al Manji A, Al Maani A, Al Rawahi B, Al Harthy K, Alyaquobi F, Al-Jardani A, Petersen E, Al Abri S (2020) COVID-19 epidemic monitoring after non-pharmaceutical interventions: the use of time-varying reproduction number in a country with a large migrant population. Int J Infect Dis 99:466–472
50. Sun J, Shi Z, Xu H (2020) Non-pharmaceutical interventions used for COVID-19 had a major impact on reducing influenza in China in 2020. J Travel Med
51. Cowling BJ, Ali ST, Ng TW, Tsang TK, Li JC, Fong MW, Liao Q, Kwan MY, Lee SL, Chiu SS, Wu JT (2020) Impact assessment of non-pharmaceutical interventions against coronavirus disease 2019 and influenza in Hong Kong: an observational study. The Lancet Public Health
52. Perra N (2021) Non-pharmaceutical interventions during the COVID-19 pandemic: a review. Phys Rep 913:1–52
53. Davies NG, Kucharski AJ, Eggo RM, Gimma A, Edmunds WJ, Jombart T, O'Reilly K, Endo A, Hellewell J, Nightingale ES, Quilty BJ (2020) Effects of non-pharmaceutical interventions on COVID-19 cases, deaths, and demand for hospital services in the UK: a modelling study. The Lancet Public Health
54. Zamir M, Shah Z, Nadeem F, Memood A, Alrabaiah H, Kumam P (2020) Non pharmaceutical interventions for optimal control of COVID-19. Comput Methods Programs Biomed 196:105642
55. Ullah S, Khan MA (2020) Modeling the impact of non-pharmaceutical interventions on the dynamics of novel coronavirus with optimal control analysis with a case study. Chaos, Solitons Fractals 139:110075
56. Panovska-Griffiths J (2020) Can mathematical modelling solve the current Covid-19 crisis?. 1–3
57. Kraemer MU, Yang CH, Gutierrez B, Wu CH, Klein B, Pigott DM, Du Plessis L, Faria NR, Li R, Hanage WP, Brownstein JS (2020) The effect of human mobility and control measures on the COVID-19 epidemic in China. Science 368(6490):493–497

58. Cartenì A, Di Francesco L, Martino M (2020) How mobility habits influenced the spread of the COVID-19 pandemic: results from the Italian case study. Sci Total Environ 741(2020):140489
59. Zhu Y, Xie J, Huang F, Cao L (2020) The mediating effect of air quality on the association between human mobility and COVID-19 infection in China. Environ Res 189:109911
60. Kermack WO, McKendrick AG (1927) A contribution to the mathematical theory of epidemics. In: Proceedings of the royal society of London. Series A, containing papers of a mathematical and physical character, vol 115(772). pp 700–721
61. Allen LJ (1994) Some discrete-time SI, SIR, and SIS epidemic models. Math Biosci 124(1):83–105
62. Brauer F, Feng Z, Castillo-Chavez C (2010) Discrete epidemic models. Math Biosci Eng 7(1):1
63. Brauer F, Castillo-Chavez C, Castillo-Chavez C (2012) In: Mathematical models in population biology and epidemiology. vol 2(40). New York, Springer, pp 508
64. Arino J, Van Den Driessche P (2003) The basic reproduction number in a multi-city compartmental epidemic model. In: Positive systems. Springer, Berlin, Heidelberg, pp 135–142
65. Zakary O, Rachik M, Elmouki I (2017) A new analysis of infection dynamics: multi-regions discrete epidemic model with an extended optimal control approach. Int J Dynam Control 5(4):1010–1019
66. Zakary O, Rachik M, Elmouki I (2017) On the analysis of a multi-regions discrete SIR epidemic model: an optimal control approach. Int J Dynam Control 5(3):917–930
67. Abouelkheir I, El Kihal F, Rachik M, Zakary O, Elmouki I (2017) A multi-regions SIRS discrete epidemic model with a travel-blocking vicinity optimal control approach on cells. J Adv Mathemat Comput Sci 20(4):1–16
68. Abouelkheir I, Rachik M, Zakary O, Elmouk I (2017) A multi-regions SIS discrete influenza pandemic model with a travel-blocking vicinity optimal control approach on cells. Am J Comput Appl Math 7(2):37–45
69. Zakary O, Rachik M, Elmouki I, Lazaiz S (2017) A multi-regions discrete-time epidemic model with a travel-blocking vicinity optimal control approach on patches. Adv Difference Equ 2017(1):120
70. Zakary O, Bidah S, Rachik M, Elmouki I (2019) Cell and patch vicinity travel restrictions in a multi-regions SI discrete epidemic control model. Int J Adv Appl Math Mech 6:30–41
71. Bidah S, Rachik M, Zakary O, Boutayeb H, Elmouki I (2018) Travel-blocking optimal control policy on borders of a chain of Regions subject to SIRS discrete epidemic model. Asian J Res Infectious Diseases 1(2):1–12
72. Chouayakh K, Rachik M, Zakary O, Elmouki I (2017) A multi-regions SEIS discrete epidemic model with a travel-blocking vicinity optimal control approach on cells. J Math Comput Sci 7(3):468–484
73. Boutayeb H, Bidah S, Zakary O, Elmouki I, Ferjouchia H, Mostafa R (2020) SIS multi-regions discrete Influenza pandemic model and travel-blocking vicinity optimal control strategy on two forms of patch. Commun Math Biol Neurosci 2020(2020):29–46
74. Tejedor S, Cervi L, Tusa F, Portales M, Zabotina M (2020) Information on the COVID-19 pandemic in daily newspapers' front pages: case study of Spain and Italy. Int J Environ Res Public Health 17(17):6330
75. Tangcharoensathien V, Calleja N, Nguyen T, Purnat T, D'Agostino M, Garcia-Saiso S, Landry M, Rashidian A, Hamilton C, AbdAllah A, Ghiga I (2020) Framework for managing the COVID-19 infodemic: methods and results of an online, crowdsourced WHO technical consultation. J Med Internet Res 22(6):e19659
76. Zarocostas J (2020) How to fight an infodemic. The Lancet 395(10225):676
77. Nielsen RK, Fletcher R, Newman N, Brennen SJ, Howard PN (2020) Navigating the 'infodemic': how people in six countries access and rate news and information about coronavirus. Reuters Institute
78. Naeem SB, Bhatti R (2020) The Covid-19 'infodemic': a new front for information professionals. Health Info Libr J 37(3):233–239

79. Alvarez-Risco A (2020) The Peru approach against the COVID-19 Infodemic: insights and strategies. The Amer J Tropical Med Hygiene tpmd200536
80. Allahverdipour H (2020) Global challenge of health communication: Infodemia in the Coronavirus disease (COVID-19) pandemic. J Educ Community Health 0–0
81. Andersen A, Godoy E (2020) Infodemia em tempos de pandemia: batalhas invisíveis com baixas imensuráveis. Revista Memorare 7(2):184–198
82. who.int/news/item/23-09-2020-managing-the-covid-19-infodemic-promoting-healthy-behaviours-and-mitigating-the-harm-from-misinformation-and-disinformation
83. Yamanis T, Nolan E, Shepler S (2016) Fears and misperceptions of the Ebola response system during the 2014–2015 outbreak in Sierra Leone. PLoS Negl Trop Dis 10(10):e0005077
84. Wigmore R (2015) Contextualising Ebola rumours from a political, historical and social perspective to understand people's perceptions of Ebola and the responses to it. Ebola Response Anthropol Platf 4
85. Cheung E (2015) An outbreak of fear, rumours and stigma: psychosocial support for the Ebola virus disease outbreak in West Africa. Intervention 13(1):70–76
86. El Kihal F, Abouelkheir I, Elmouki I (2023) What has gone around may come back around: reinfection in the extended stochastic multi-region control of infodemics and seasonal coronaviruses. Commun Math Biol Neurosci 2023(1):125–146
87. Zhang Z, Zeb A, Hussain S, Alzahrani E (2020) Dynamics of COVID-19 mathematical model with stochastic perturbation. Adv Difference Equ 2020(1):1–12
88. Rihan FA, Alsakaji HJ, Rajivganthi C (2020) Stochastic SIRC epidemic model with time-delay for COVID-19. Adv Difference Equ 2020(1):1–20
89. Sepulveda-Salcedo LS, Vasilieva O, Svinin M (2020) Optimal control of dengue epidemic outbreaks under limited resources. Stud Appl Math 144(2):185–212
90. Zhang H, Yang Z, Pawelek KA, Liu S (2020) Optimal control strategies for a two-group epidemic model with vaccination-resource constraints. Appl Math Comput 371:124956
91. Dondorp AM, Hayat M, Aryal D, Beane A, Schultz MJ (2020) Respiratory support in COVID-19 patients, with a focus on resource-limited settings. Am J Trop Med Hyg 102(6):1191
92. Shammi M, Bodrud-Doza M, Islam ARMT, Rahman MM (2020) COVID-19 pandemic, socioeconomic crisis and human stress in resource-limited settings: a case from Bangladesh. Heliyon 6(5):e04063
93. Bartsch SM, Ferguson MC, McKinnell JA, O'shea KJ, Wedlock PT, Siegmund SS, Lee BY (2020) The potential health care costs and resource use associated with COVID-19 in The United States: a simulation estimate of the direct medical costs and health care resource use associated with COVID-19 infections in the United States. Health Aff 39(6):927–935
94. Cepiku D, Marchese B, Mastrodascio M (2021) The Italian response to the economic and health crises: a budgetary comparison. J Public Budg Account Financ Manag 33(1):47–55
95. Zia MM, Kalim I (2021) COVID-19 and its socio-economic and psychological impacts on Italy. J European Stud (JES) 37(1):92–108
96. Hamdache A, Saadi S, Elmouki I, Zouhri S (2013) Two therapeutic approaches for the treatment of HIV infection in AIDS stage. J Appl Mathemat Sci 7(105):5243–5257
97. Zouhri S, Saadi S, Elmouki I, Hamdache A, Rachik M (2013) Mixed immunotherapy and chemotherapy of tumors: optimal control approach. Int J Comput Sci Issues (IJCSI) 10(4):81
98. Zakary O, Rachik M, Elmouki I (2015) On effectiveness of an optimal antiviral bitherapy in HBV-HDV coinfection model. Int J Comput Appl 127(12):1–10
99. Elmouki I, Saadi S (2016) Quadratic and linear controls developing an optimal treatment for the use of BCG immunotherapy in superficial bladder cancer. Optimal Control Appl Methods 37(1):176–189
100. Hamdache A, Saadi S, Elmouki I (2016) Free terminal time optimal control problem for the treatment of HIV infection. An Int J Optimizat Control: Theories and Appl (IJOCTA) 6(1):33–51

101. Zakary O, Rachik M, Elmouki I (2017) How much time is sufficient for benefiting of awareness programs in epidemics prevention? A free final time optimal control approach. Int J Adv Appl Math Mech 4:26–40
102. Chouayakh K, Rachik M, Satori K, Bekkali CE, Elmouki I (2017) Trophic and halieutic dynamics of grazer–predator fishes: harvesting optimal control policies for the environmental sustainability and bioeconomic cases. Modeling Earth Syst Environ 3:567–580
103. Alkama M, Larrache A, Rachik M, Elmouki I (2018) Optimal duration and dosage of BCG intravesical immunotherapy: a free final time optimal control approach. Mathemat Methods in the Appl Sci 41(5):2209–2219
104. Aboulaich R, Darouichi A, Elmouki I, Jraifi A (2017) A stochastic optimal control model for BCG immunotherapy in superficial bladder cancer. Mathemat Modelling of Natural Phenomena 12(5):99–119
105. Abouelkheir I, El Kihal F, Rachik M, Elmouki I (2019) Optimal impulse vaccination approach for an SIR control model with short-term immunity. Mathematics 7(5):420
106. Leiva H (2023) Pontryagin's maximum principle for optimal control problems governed by nonlinear impulsive differential equations. J Mathem Appl 46(1):15–68
107. Abouelkheir I, El Kihal F, Elmouki I (2024) Multi-region optimal impulse control through example of immunization policy at large geographical scale with discussion of the closure case. Commun Math Biol Neurosci
108. Elmouki I, Saadi S (2016) BCG immunotherapy optimization on an isoperimetric optimal control problem for the treatment of superficial bladder cancer. Int J Dynam Control 4:339–345
109. Hamdache A, Elmouki I, Saadi S (2014) Optimal control with an isoperimetric constraint applied to cancer immunotherapy. Int J Comput Appl 94(15):31–37
110. Alkama M, Rachik M, Elmouki I (2017) A discrete isoperimetric optimal control approach for BCG immunotherapy in superficial bladder cancer: discussions on results of different optimal doses. Int J Appl Computat Mathem 3:1–18
111. Dar M, Swamy L, Gavin D, Theodore A (2021) Mechanical-ventilation supply and options for the COVID-19 pandemic. Leveraging all available resources for a limited resource in a crisis. Annals of the American Thoracic Soc 18(3):408–416

Inequality of Clean Water in Africa

8

Daniel Jinnefält

Abstract

This chapter delves into the origins, repercussions, and potential remedies for the disparity in access to clean water in Africa. It scrutinizes the influence of climate change, inadequate governance, population expansion, and urbanization in Africa, which collectively contribute to an escalating demand for water. The examination extends to the gender bias faced by women and children due to this heightened demand, resulting in reduced access to time for income generation and education for females. Overall, unequal access to clean water leads to insufficient nutritional intake and health challenges. Moreover, the environment suffers significantly from water scarcity and fluctuations in water availability, sparking conflicts in areas with limited water resources. Traditional solutions, such as desalination and borehole drilling, are commonly advocated for addressing water shortages, but these measures tend to exacerbate inequalities in water access and harm the environment. Contrarily, this chapter advocates for the viability of water harvesting and conservation technologies, which are economically feasible, locally accessible, and environmentally sustainable. These solutions can be implemented on either a small-scale or a national level. Notably, Africa experiences considerable precipitation, and harnessing this resource effectively can either pose challenges, such as water erosion, or be a valuable asset if managed wisely in agriculture, road networks, and industry.

Keywords

Water • Inequality • Environment • Sustainable • Water harvesting • Gender • Erosion

D. Jinnefält (✉)
MediTec Group, Swiss Tower, 1401-02, JLT, Dubai, UAE
e-mail: ceo@meditec.seg

B. W. K. Son (ed.), *The Landscape of Global Health Inequity*, Integrated Science 22,
https://doi.org/10.1007/978-3-031-60502-4_8

1 Introduction

Water, the lifeblood of our existence, is central to health, hygiene, and a dignified life. Recognizing this, the United Nations' Sustainable Development Goals prioritize ensuring everyone has access to clean water and sanitation. However, in Africa, millions grapple with water scarcity, facing a daily struggle for survival. This crisis is not solely about a lack of resources but also about unequal access and management [1].

This chapter delves into the root causes and far-reaching consequences of Africa's unequal water distribution and management. The chapter explores the factors contributing to this disparity, ranging from climate change and population growth to inadequate infrastructure and governance. The consequences are stark: widespread disease, poverty, and even conflict. By understanding the complex dynamics of this crisis, we can forge effective solutions. This chapter will propose possibilities for ensuring equitable access to clean water for all Africans, paving the way for a healthier, more prosperous, and sustainable future for the continent.

2 Causes of Water Inequality in Africa

The African people's water inequality has many causes; however, there are few solutions due to the environmental impact caused by the engineering methods of handling water shortages. The solution for addressing the inequality of water access must be sought through conservational techniques and proper management of water resources. Water inequality in Africa has multiple and interrelated causes, such as climate change, poor governance, population growth, urbanization, and gender discrimination (Table 1).

Table 1 Causes of water inequality in Africa

Climate change	– Makes the continent vulnerable to extreme events – Exacerbates land degradation and desertification – Has a significant impact on agricultural productivity
Poor governance	– Contributes to water inequality – Corruption mismanagement hinders equal water access – Fuel conflicts over water resources
Population growth	– Population to double by 2050 to 2.5 Billion people – Need for water infrastructure and services improvement – 400 million people in sub-Saharan Africa lack water access
Urbanization	– Urban areas overused for water supply and sanitation—Urban areas overused for water supply and sanitation – Informal/illegal vendors deliver contaminated water

Source author

2.1 Climate Change

Climate change affects water availability and quality in Africa as it alters rainfall patterns, temperature, and evaporation patterns. According to the World Resources Institute [2], Africa faces some of the highest water risks in the world. The continent is vulnerable to extreme events such as droughts and floods, which can reduce water supply, damage infrastructure, and contaminate sources. The water distribution through rain precipitation is unevenly divided over the African continent, where the Sahel predominantly has lower rainfall and a high variability over the year. Moreover, climate change can exacerbate desertification and land degradation due to human activities or natural factors. Desertification reduces the amount of arable land and freshwater resources, especially in arid and semi-arid regions [3].

Climate conditions can have a significant impact on agricultural productivity. Tropical agriculture, for example, is up to 50% less productive than agriculture in temperate regions [4]. The disparity is linked to the inadequacy of tropical soils for large-scale agriculture and the rising prevalence of diseases affecting both humans and plants. The specific impacts of climate on agriculture vary based on rainfall patterns. For instance, low deserts receive less rainfall than the widely accepted threshold of 500 millimeters/year required for rain-fed agriculture but do not face the same disease-related challenges. In light of these observations, climates less conducive to agriculture are correlated with lower agricultural returns, potentially heightening the probability of conflict, assuming all other factors remain constant [5].

2.2 Poor Governance

Poor governance is another major factor contributing to water inequality in Africa. Numerous studies have shown a solid and lasting relationship between a country's governance quality and overall resource management [6]. Many African countries need more effective policies, institutions, and regulations to manage their water resources and services. Corruption, mismanagement, and lack of accountability hinder the provision of adequate and equitable water access. Poor governance can fuel conflicts over water resources as different groups compete for scarce and valuable water sources. According to the United Nations Environment Programme, water-related conflicts have increased by 270% in Africa in the last decade [7].

2.3 Population Growth

Population growth is also putting pressure on the existing water resources in Africa. The continent's population is projected to double by 2050, reaching 2.5 billion people. Hence, more people will need more water for drinking, sanitation, agriculture, industry, and energy. However, the current water infrastructure and services must be improved to meet the expanding demand where, according to the World Health Organization, 400 million people in sub-Saharan Africa lack access to essential drinking water services [8].

2.4 Urbanization

Urbanization is another demographic trend that affects water inequality in Africa. As of 2021, an estimated 609 million people in Africa live in urban areas. The urban population has been growing annually and is forecast to reach 722 million by 2026, which will constitute over 5% urban growth in five years [9]. More people are moving from rural to urban areas for better opportunities and services [10]. However, many metropolitan areas cannot cope with the rapid influx of people and provide them with sufficient and safe water supply and sanitation. As a result, many urban dwellers rely on informal or illegal sources of water, such as vendors, wells, or streams, which are often contaminated or expensive [11].

3 Consequences of Water Inequality in Africa

Water inequality in Africa has severe and far-reaching consequences for human development, health, environment, and peace. Water inequality hampers human development in Africa by limiting people's opportunities and capabilities to achieve their full potential. Lack of access to clean water affects various aspects of human well-being, such as education, income, nutrition, and gender equality [12] (Table 2).

3.1 Gender and Education

Gender discrimination is also a cause and a consequence of water inequality in Africa. Women and girls are disproportionately affected by the lack of access to clean water, as they collect water for their households. In sub-Saharan Africa, more than 25% of the population must walk 30 min or more to collect water. This task takes time away from education, income generation, child care, and household chores. Moreover, women and girls face health risks and violence when they walk long distances or use unsafe facilities to fetch water or relieve themselves [13]. Lack of access to clean water affects the education outcomes of children in Africa. Children who walk long distances or miss school days to collect water have less time and energy to study or attend classes. Moreover, children who drink unsafe water or lack proper sanitation facilities at school are more likely to suffer from illnesses or infections that impair their learning abilities or cause absenteeism [14].

Table 2 Consequences of water inequality in Africa

Gender and education	– Women and girls must collect water for households
	– 25% must walk 30 min or more to collect water
	– Affects education, income generation, and health risks
	– Loss of access to education and sanitary illnesses impairing learning abilities
Income	– Water scarcity impairs the profitability of agriculture
	– Water scarcity limits industry, tourism, and energy
Nutrition	– Water scarcity affects the food availability and quality
	– Water scarcity leads to food insecurity and malnutrition
Health	– Water inequality harms the health of millions of people
	– 115 people die daily by diseases due to water inequality
Environment	– Affects flor/fauna biodiversity and climate regulation
	– Harmful substances, chemicals, nutrients, pathogens
Peace	– Water conflicts occur at international and local levels
	– Impact human lives, livelihoods, and stability

Source author
Respective consequences are discussed below

3.2 Income

Lack of access to clean water affects the income opportunities of people in Africa. Water scarcity and pollution reduce the productivity and profitability of agriculture, which is the primary source of livelihood for 90% of sub-Saharan Africa's rural population [15]. Moreover, lack of access to clean water limits the development of other sectors, such as industry, tourism, or energy, which require large amounts of water for their operations [16].

3.3 Nutrition and Health

The lack of access to clean water affects the nutrition status of people in Africa. Water scarcity and contamination affect the availability and quality of food crops, which can lead to food insecurity and malnutrition [17]. Water inequality jeopardizes the health of millions of people in Africa by exposing them to water-related diseases and deaths. According to the UNEP, 115 people die daily from diseases linked to poor sanitation, poor hygiene, or contaminated water in Africa [18].

3.4 Environment

Water inequality harms the environment in Africa by degrading the natural resources and ecosystems that sustain life on Earth. Water scarcity and pollution affect the quality and quantity of freshwater resources, such as rivers, lakes, wetlands, and groundwater. These resources provide vital services for humans and nature, such as drinking water for wild fauna, biodiversity, climate regulation, and flora. For example, in South Africa, Uganda, and Zambia, the gap between rural and urban water services persists, affecting both people and ecosystems [19]. Human activities such as deforestation, agriculture, industry, mining, and urbanization threaten the integrity and functioning of these resources by altering their flow regimes, reducing their recharge rates, increasing their sedimentation and salinization levels, and introducing harmful substances such as chemicals, nutrients, and pathogens [20].

3.5 Conflicts

Water inequality threatens peace and security in Africa by creating or exacerbating conflicts over water resources. For example, in the Turkana region in northern Kenya, droughts have been recurring annually since the 1990s. In the 2020s, over 400,000 livestock perished. As livestock is the lifeline of pastoralists in this region, these annual ordeals of drought are intensifying armed conflicts in the region [21]. Water conflicts can occur at different levels, such as between countries, within countries, or communities. Water conflicts negatively impact human lives, livelihoods, and stability [22]. Various factors, such as competing demands for water, unequal access to water, poor management of water resources, or environmental degradation of water sources, can trigger water conflicts; where for example, in the Turkana area, 60% of the population faces acute food insecurity and hunger [23].

4 Challenges and Solutions for Water Inequality in Africa

The general contemporary methods of encountering water shortage in Africa consist of desalination and drilling boreholes as water harvesting, and conservation is a method that has been used over centuries and is gaining momentum lately [24] (Table 3).

4.1 Desalination

Desalination is a method that is gaining more interest in Africa today as a means of handling water shortages. South Africa has today plants in Cape Town, where

Table 3 Challenges and solutions for water inequality in Africa

Desalination	– Raising questions about the remaining water and salt
	– May harm marine life and coastal ecosystems
	– Would need massive investments in transport systems
	– To reach the inner of the African continent
Boreholes	– Exacerbate inequality, as people without the means to drill
	– Boreholes will have no access to water
	– Boreholes deeper than 60 m lower the water table below the reach of plant roots, leading to vegetation loss and reduced access to water for plants and animals
	– Borehole water has caused the Earth's rotational pole to wander, causing more than 500 mm of sea level rise
Water harvesting and conservation	– Crucial for agriculture expansion, reducing soil erosion
	– Ensure communities have equitable access to clean water
	– Recycling involves treating wastewater for other purposes, such as irrigation or industrial use

Source author
These challenges and respective solutions are subsequently discussed below

a temporary desalination plant produces 2000 cubic meters of water per day, soon to increase to 7000 using reverse osmosis technology as more ambitious plans are in the pipe to meet the water demands of its 4 million residents [25]. Ongoing projects in Africa are setting up desalination plants in Algeria, Ghana, Namibia, Kenyan Tanzania, Somalia, and Madagascar [26].

Case studies collecting data from Australia, Israel, and the USA show how the desalination process involves removing salt from seawater, raising questions about the structure for distribution as the PPP usage rises with the technology introduction and as governments are dependent on private sector partners for the design and operations [27]. The byproduct of desalination is a hypersaline slurry known as brine, which poses challenges in disposal due to its high salt concentration and density [28]. Brine waste is commonly disposed of in the ocean, creating environmental concerns; where to mitigate these impacts, various strategies can be employed. Case studies have been conducted on this technology in Israel, where one approach is to mix brine with other water sources to dilute its salinity before discharge. Additionally, brine can be discharged through diffusers, which promote brine mixing with ocean water far from the shore at depth [29]. However, most desalination facilities do not have the Israel Case study technical capacity, and it is crucial to consider the potential environmental impacts of desalination.

Most desalination processes are energy-intensive, increasing dependence on fossil fuels. Failure to adopt renewable energy sources for freshwater production during desalination is thought to exacerbate the issue of further oil dependency

[30]. It is essential to consider the disposal of brine waste and the overall environmental impacts of desalination. The high salt concentration and density of brine waste require careful management to prevent harm to marine life and coastal ecosystems. Additionally, the energy-intensive nature of desalination processes necessitates using renewable energy sources to mitigate potential contributions to climate change.

Proper assessment and implementation of desalination projects are crucial to minimize environmental impacts and promote sustainable freshwater production [31]. Furthermore, desalination would need massive investments in transport systems to reach the inner of the African continent, where considerate parts of the African continent's demography are concentrated; why costs for these transports would have to be added, whereby the inequality for water access and affordability would be worsening.

4.2 Boreholes

Drilling boreholes is an even more utilized engineering method of extracting water otherwise unavailable for human consumption. However, this method of accessing water has multiple environmental and health impacts. Drilling boreholes deeper than 60 m lowers the water table below the reach of plant roots, leading to vegetation loss and reduced access to water for animals. This exacerbates inequality further, as people without the means to drill boreholes will have no access to water.

The lack of water for flora and fauna can also negatively impact the general population, which relies on shallow water for their livelihood and the welfare of their animals. The depletion of shallow water makes typical flora used for medicinal, food, and firewood purposes scarce, further widening the inequality in quality and life expectancy [32].

Furthermore, people in the area who do not have access to drilled boreholes will suffer from the disparity of shallow water and wells drying out [33]. However, even for people with access to boreholes, this method of retrieving water carries many negative issues, as water from deep aquifers can have high levels of minerals and gasses, such as fluoride and iron [34]. These minerals can harm human health, especially when consumed unfiltered, as the water from very deep aquifers often has a mineral composition that is very different from shallow water.

Moreover, drilling boreholes is done with little governance oversight, resulting in abundant boreholes made in certain areas where the population may afford them. In poverty areas, it is typical for non-governmental organizations to sink boreholes to help the people in the rural districts, as this is not done with the governance needed to understand the sustainability effects of the drilling [35].

According to a study published in Geophysical Research Letters, humans have shifted such a large mass of water by pumping it out of the ground and moving it elsewhere that the Earth tilted nearly 80 centimeters east between 1993 and 2010 alone. Overusing borehole water has caused the Earth's rotational pole to wander almost a meter in two decades, leading to changes in the planet's rotational pole. The redistribution of groundwater has been found to have the most immense impact on the drift of the rotational pole among climate-related causes. The study estimates that humans pumped 2150 gigatons of groundwater from 1993 to 2010, which has caused more than half a centimeter of sea level rise [36].

Following the understanding of the solutions used today for handling inequality of water and water scarcity, it is evident that proper management of groundwater resources is crucial to minimize environmental impacts and promote sustainable freshwater production. Improving governance is essential for ensuring the equitable and sustainable management of water resources and services in Africa [37].

4.3 Water Harvesting and Conservation

Water harvesting and conservation practices are crucial in reducing soil erosion and enhancing water conservation, which is essential for addressing African water inequality. By implementing effective water conservation strategies, we can ensure equitable access to clean water resources for all communities, promoting sustainable development and improving the lives of millions.

Governance structural improvement that involves and strengthens the policies, institutions, and regulations that govern the allocation, distribution, and protection of water resources and services is necessary. Moreover, it involves enhancing the participation, accountability, and transparency of all stakeholders involved in water management, such as governments, the private sector, civil society, and local communities [38]. Furthermore, investing in water infrastructure is crucial for improving the access and quality of water supply and sanitation in Africa, which involves expanding and upgrading the physical infrastructure that delivers water services to people, such as harvesting structures, pipes, tanks, filters, or for sanitary purpose latrines. Moreover, it involves investing in innovative technologies that can increase the efficiency and reliability of water services, such as rainwater harvesting systems, possibly using mobile applications to understand how to harvest available water resources best.

Hydrogeologic studies over the past decades show that rainwater amounts have ncreased significantly in many parts of Africa from 1980 to 2024 [39]. In 2024, the rainwater precipitation was much higher than in 1980 and will continue to increase in the foreseeable future, with models looking up until 2060 [40]. Mean annual precipitation increased by 0.6 mm/year with a 5% rise of mean annual rainfall in Sub-Saharan Africa from 1980 to 2020, with a predicted further continuation of this rainfall rise by 2060, according to NCAR-PCM3 [41]. Simultaneously, the surface temperature and rainfall variability have increased, further calling for better governance of water resources to benefit human consumption, natural resources, industry, and agricultural use [42, 43].

Agriculture is a major water consumer, and improving irrigation efficiency is crucial for water conservation, which can be achieved through drip irrigation, precision irrigation, scheduling irrigation based on crop needs, and rainwater harvesting, which involves collecting and storing rainwater for later use [44]. This practice is precious in areas with limited access to other water sources [45]. Watershed management involves protecting and restoring watersheds, which are the areas of land that drain into a particular body of water. Protecting watersheds can reduce soil erosion, improve water quality, and enhance water storage [46]. Water reuse and recycling involve treating wastewater for other purposes, such as irrigation or industrial use. This practice reduces the demand for freshwater and helps to conserve water resources. Raising public awareness about water conservation is essential for promoting behavioral changes. Education programs can teach individuals and communities about the importance of water conservation and provide practical tips for reducing water consumption [47].

Several successful water conservation projects are being implemented across Africa, demonstrating the effectiveness of these practices in addressing water scarcity and inequality. The African Development Bank (AfDB) provides funding and technical support for African water conservation projects. The AfDB's projects focus on improving access to water and sanitation, promoting sustainable water management practices, and addressing water inequality [48]. These projects have shown a capacity for implementation of the use of existing rainwater as a sole source of water. Implementing water harvesting methods that effectively collect and store rainwater saves the soil from devastation and boosts agricultural capacity due to constant water access over the year [49].

5 Conclusion

The need for water in Africa is rising with the emerging population growth, industrialization, and expanding agricultural industry needs. The capacity to meet this need by using technical solutions such as desalination and piping of water into the African continent or drilling deeper boreholes has proven futile and even extensively damaging to the local, regional, and global environment.

For Africa to see the inequality of water dispersal decrease, the situation needs a hands-on and local solution that can be scaled up on a national and regional level, whereby the rising levels of rainwater on the continent are strategically harvested. This way, water is kept from harming the environment, the livelihood of people, and agriculture. It is instead used in a manner where the water in industry and agriculture is conservatively managed using fourth industrial revolution means, such as AI-triggered robots for subsurface drip irrigation and drone crop control, industries where constant flow technologies are introduced to maintain and elevate the production capacities with limited needs for water usage and damage, plus introduction of water reuse in industries whereby water that has been used in the processes may be cleaned up and used for lower grade functions. The water from industry and agriculture must be revitalized and reused in other parts of society, such as the irrigation of public greenhouses and parks. Fundamentally, there needs to be a shift from trying to take out water that is not available without extreme environmental consequences to focusing on harvesting the rainwater that is coming in abundance and using this water wisely through good governance instead of allowing this rainwater to destroy livelihoods, agricultural land, and infrastructure to rather enabling rainwater to improve the financial output from agriculture and industry.

References

1. United Nations (2022) Sustainable Development Goal 6: Ensure availability and sustainable management of water and sanitation for all. [Online]. Available at: https://sdgs.un.org/goals/goal6
2. World Resources Institute (WRI) (2022) Land degradation and desertification: what you need to know. https://www.who.int/news-room/questions-and-answers/item/climate-change-land-degradation-and-desertification
3. Shiri J, Yaseen Z, Kim S, Il-Moon C, Nourani V, Zounemat-Kermani M (2021) Development of artificial intelligence models for well groundwater quality simulation: different modeling scenarios. PLoS ONE 16(5):e0251510
4. Gallup JL, Sachs JD, Mellinger AD (1999) Geography and economic development. Int Reg Sci Rev 22(2):179–232
5. Organisation for Economic Co-operation and Development (2009) Agriculture and climate change: impacts, mitigation and adaptation. OECD Publishing. https://doi.org/10.1787/9789264086876-en
6. Phillips SD (2019) Proximities of violence: civil order beyond governance institutions. Int Stud Quart. https://doi.org/10.1093/isq/sqz036
7. Sulton A (2021) When? Jackson Advocate 83(25):8A
8. UNICEF/WHO (2022) Africa to drastically accelerate progress on water, sanitation and hygiene—report. UNICEF. https://www.unicef.org/wca/press-releases/africa-drastically-accelerate-progress-water-sanitation-and-hygiene-report
9. WHO (2021) Progress on household drinking water, sanitation, and hygiene 2000–2020: Five years into the SDGs. World Health Organization (WHO) and the United Nations Children's Fund (UNICEF), Geneva
10. Statista (2023) Urban population in Africa 2000–2026. Retrieved from Statista website
11. Manungufala T (2021) Water scarcity: classification, measurement and management. In: Encyclopedia of the UN sustainable development goals, pp 1–14. SpringerLink
12. Evaristo J, Jameel Y, Tortajada C, Wang RY, Horne J, Neukrug H, David CP, Fasnacht AM, Ziegler AD, Biswas A (2023) Water woes: the institutional challenges in achieving SDG 6. Sustain Earth Rev 6, Article number: 13
13. UNICEF (2022) Africa to drastically accelerate progress on water, sanitation and hygiene—report. [Press release]. Dakar, Senegal: UNICEF Senegal. https://www.unicef.org/senegal/en/press-releases/africa-drastically-accelerate-progress-water-sanitation-and-hygiene-report
14. UNICEF (2016) Collecting water is often a colossal waste of time for women and girls. UNICEF. https://www.unicef.org/press-releases/unicef-collecting-water-often-colossal-waste-time-women-and-girls
15. Matchawe C, Bonny P, Yandang G, Mafo HCY, Nsawir BJ (2022) Water shortages: cause of water safety in sub-Saharan Africa. In: Drought—impacts and management, p 284. IntechOpen. https://doi.org/10.5772/intechopen.103927
16. Mao B (2022) Work together to improve access to clean water worldwide. World Economic Forum1
17. Oskam MJ, Pavlova M, Hongoro C, Groot W (2021) Socio-economic inequalities in access to drinking water among inhabitants of informal settlements in South Africa. Int J Environ

Res Public Health 18(19):10528. https://doi.org/10.3390/ijerph181910528. PMID: 34639828; PMCID: PMC8507892
18. UNEP (2020) Fecal sludge management in Africa: socio-economic aspects, human and environmental health implications. Accessed Dec 8 2023. https://wedocs.unep.org/bitstream/handle/20.500.11822/34350/FSM.pdf?sequence=1&isAllowed=y
19. Chitonge H, Mokoena A, Kongo M (2020) Water and sanitation inequality in Africa: challenges for SDG 6. In: Ramutsindela M, Mickler D (eds) Africa and the sustainable development goals. Sustainable development goals series. Springer, Cham. https://doi.org/10.1007/978-3-030-14857-7_20
20. Sale FA, Agbidye F (2011) Impact of human activities on the forest and their effects on climate change. Aust J Basic Appl Sci 5:863–867
21. Turner MD, Schlecht E (2019) Livestock mobility in sub-Saharan Africa: a critical review. Pastoralism 9, Article number: 131
22. Hendrix CS, Glaser SM (2007) Trends and triggers: Climate, climate change and civil conflict in Sub-Saharan Africa. Polit Geogr 26(6):695–715. https://doi.org/10.1016/j.polgeo.2007.06.006
23. ACAPS (2022) Impact of drought: Turkana and Marsabit counties. Retrieved from https://www.acaps.org/fileadmin/Data_Product/Main_media/20221207_acaps_thematic_report_horn_of_africa_impact_of_drought_in_kenya_0.pdf
24. Manungufala T (2021) The general contemporary methods of encountering water shortage in Africa consist of desalination and drilling boreholes as water harvesting, and conservation is a method that has been used over centuries and is gaining momentum lately. In: water scarcity: classification, measurement and management, pp 1–14. SpringerLink
25. The Conversation (2018) Desalination: global examples show how Cape Town could up its game. The Conversation
26. Fluence News Team (2022) Is desalination a key to Africa's water future? Retrieved from Fluence. https://www.fluencecorp.com/is-desalination-key-to-water-future-of-africa/
27. Greer RA, Lee K, Fencl A et al (2021) Public-private partnerships in the water sector: the case of desalination. Water Resour Manage 35:3497–3511. https://doi.org/10.1007/s11269-021-02900-9
28. United Nations Environment Programme (2021) Five things to know about desalination. UNEP
29. World Bank (2019) The role of desalination in an increasingly water-scarce world. P. 89. World Bank, Washington, DC. https://www.unep.org/news-and-stories/story/five-things-know-about-desalination
30. Al-Madani H, Al-Sulaiman F, Al-Senafy M (2021) Renewable energy sources for desalination: a review. J Water Process Eng 44:102276. https://doi.org/10.1016/j.jwpe.2021.102276
31. A review of the management and treatment of brine solutions (2017) Royal Society of Chemistry. https://pubs.rsc.org/en/content/articlelanding/2017/ew/c6ew00339g
32. Soulsby D (2015) Borehole drilling and rehabilitation under field conditions. Practical Action Publishing
33. Ritchie H, Spooner F, Roser M (2019) Clean water. Our world in data. https://ourworldindata.org/clean-water
34. Quality of Ground Water—USGS Publications Warehouse (n.d.) Accessed Dec 2023 from https://pubs.usgs.gov/gip/gw/quality.html
35. Water Inequality. (n.d.). National Geographic Society. Accessed December 8, 2023 from https://www.nationalgeographic.org/article/water-inequality/
36. Seo KW et al (2023) Groundwater extraction and its impact on Earth's rotational pole. Geophys Res Lett 50(12):1–9. https://doi.org/10.1029/2023GL098765
37. Unpacking Water Governance: A Framework for Practitioners (2023) Water|Free Full-Text 12(3):827. https://www.mdpi.com/2073-4441/12/3/827
38. Johnston M (2003) Good governance: rule of law, transparency, and accountability. UNESCO
39. Hijmans RJ, Cameron SE, Parra JL, Jones PG, Jarvis A (2005) Very high resolution interpolated climate surfaces for global land areas. Int J Climatol 25:1965–1978

40. Jones PG, Thornton PK (2013) Generating downscaled weather data from a suite of climate models for agricultural modelling applications. Agric Syst 114:1–5
41. Kostandini G, Tanellari E, Gaskell J (2021) The effect of land tenure and erosion measures on productivity and investments: plot and household level evidence from Mali. J Develop Areas 55(2):365–385
42. The World Bank (2023) Average precipitation in depth (mm per year)—Africa. Food and Agriculture Organization, electronic files and web site. https://data.worldbank.org/indicator/AG.LND.PRCP.MM?locations=A9&view=map
43. World Meteorological Organization (2020) State of the Climate in Africa: 2019; WMO—No. 1253 World Meteorological Organization: Geneva, Switzerland
44. Daukla S, Knowles, JM, Shrestha NK (2014) How to determine run time and irrigation cycles for drip irrigation: tomatoes and pepper examples. Bulletin AE500, University of Florida Extension
45. Wu H, Sun L, Liu Z (2022) Ecosystem service assessment of soil and water conservation based on scenario analysis in a hilly red-soil catchment of southern China. Water 14(8):1284. https://doi.org/10.3390/w14081284
46. Kesari KK, Soni R, Jamal QMS, Tripathi P, Lal JA, Jha NK, Siddiqui MH, Kumar P, Tripathi V, Ruokolainen J (2021) Wastewater treatment and reuse: a review of its applications and health implications. Water Air Soil Pollut 232(5):208. https://doi.org/10.1007/s11270-021-05154-8
47. Raising public awareness (C8.02)—GWP (n.d.) Global water partnership. Accessed Dec 2023, from https://www.gwp.org/en/learn/iwrm-toolbox/Management-Instruments/Promoting_Social_Change/Raising_public_awareness/
48. African Development Bank Group (n.d.) Water supply and sanitation. https://www.afdb.org/en/topics-and-sectors/sectors/water-supply-sanitation
49. Food and Agriculture Organization of the United Nations (2019) Soil and water conservation practices to reduce soil erosion and enhance water conservation. [Online]. https://www.fao.org/family-farming/detail/zh/c/1618121/

Exploring Global Dietary Data Gaps in Relationship to Nutrition Inequity: A Case of Sodium Intake

9

Yutong Chen, Bingjie Zhou, and Elena N. Naumova

Abstract

Adequate dietary intake plays an important role in fostering growth and preventing diseases. Understanding dietary needs at both local and global levels provides valuable insights for various stakeholders, including researchers, policymakers, public health practitioners, and healthcare professionals, guiding informed dietary decisions. In this chapter, we explore the critical role of adequate dietary intake in promoting human health and preventing diseases, with a special emphasis on sodium intake as a key dietary component. Despite the global recognition of a healthy diet as fundamental to well-being, achieving optimal nutrition remains challenging due to various factors, such as food insecurity, unhealthy lifestyle behaviors, and environmental changes. Moreover, information on dietary intake is limited, inconsistent, incomplete, or outdated. We detected substantial discrepancies in existing global dietary data sources, revealing knowledge gaps in terms of dietary data collection and factors contributing to the availability, completeness, inconclusiveness, consistency, and up-to-date accuracy of nutrition intake data. Specifically, we focused on two levels of databases: individual-level dietary intakes from the National Health and Nutrition Examination Survey versus the China Health and Nutrition Survey, and population-level intakes from the Global Dietary Databases versus the Global Burden of Disease. These databases are among the most complete and comprehensive repositories of relatively granular intake estimates.

Y. Chen · B. Zhou · E. N. Naumova (✉)
Division of Nutrition Epidemiology and Data Science, Friedman School of Nutrition Science and Policy, Tufts University, Boston, MA, USA
e-mail: Elena.Naumova@tufts.edu

Y. Chen
e-mail: Yutong.Chen@tufts.edu

B. Zhou
e-mail: Bingjie.Zhou@tufts.edu

B. W. K. Son (ed.), *The Landscape of Global Health Inequity*, Integrated Science 22,
https://doi.org/10.1007/978-3-031-60502-4_9

As we systematically compared the content of nutrition databases, we highlighted populations for whom dietary records are sparse and countries with limited resources to conduct nutrition monitoring and adhere to established reference levels and recommended thresholds for key food groups and nutrients. Throughout this chapter, we offered a deeper understanding of the essential roles the global data repositories play in shaping our comprehension of global nutrition dynamics. By investigating the intricate dimensions of dietary data, we underscored the urgent need for standardization and harmonization of data collection and analysis. These improvements are crucial to bridge gaps, reveal existing disparities, and provide insights into opportunities for enhancing nutrition surveillance and public health efforts on a global scale, thereby addressing the pressing issue of global health inequity.

Keywords

Nutrition surveillance • Health equity • Global burden disease • Global dietary database • Sodium consumption

1 Introduction

Nutrition, recognized as a major determinant of human health, also functions as an important indicator of health equity [1]. Consequently, monitoring dietary intakes is imperative, offering valuable insights into the dietary patterns and habits of populations [1, 2]. This information serves as a reference for formulating effective public health strategies aimed at enhancing overall wellness and reducing the risk of chronic diseases [2–4]. Nutrition surveillance, defined as a system established to continuously monitor the dietary intake and nutritional status of a population or selected population groups using a variety of data collection methods whose ultimate goal is to lead to policy formulation and action planning, serves as an indispensable strategy for collecting pertinent information [3, 5].

In addition to the importance of monitoring dietary intake, understanding the global burden of disease is essential for achieving health equity. Health disparities, determined by socioeconomic factors, geographical locations, and access to resources, significantly influence the burden of disease within populations, especially the disadvantaged populations [6, 7]. These disparities become clear in the prevalence of nutrition-related chronic diseases, which also tend to affect more underprivileged communities [6, 8]. By examining the links between dietary intake and health outcomes, we can gain a comprehensive understanding of the uneven distribution of health burdens [9–11]. This insight is essential for developing tailored public health initiatives that aim to reduce these disparities and promote health equity [8, 11]. The integration of the burden of disease analysis with nutrition surveillance data provides a powerful tool for identifying at-risk populations and developing effective strategies to address the root causes of health inequities [12–15]. This comprehensive approach ensures that public health policies and intervention programs are not only effective but also equitable, ultimately leading to a healthier global population where access to adequate nutrition and health resources is a universal right, not a privilege [11–15].

The comparison of dietary intakes between populations and across countries is a crucial aspect of global nutrition research. Such comparison helps researchers understand the diverse dietary patterns and behaviors, indicating the nutritional strengths and vulnerabilities of various groups and countries [16, 17]. By analyzing the differences, researchers can identify unique nutritional needs and potential public health risks specific to a population or region. However, the comparison requires a standard data structure to ensure comparability. Harmonizing methodologies and metrics for dietary intake assessment across studies and countries becomes a necessary step for generating accurate insights. This standardization not only facilitates cross-country comparisons but also enhances the reliability and validity of global dietary data. Researchers also need to pay attention to data completeness and structural missingness that may lead to biased, erroneous, or misleading inferences [18–20]. Through this process, it becomes possible to develop targeted, culturally sensitive nutrition intervention strategies and policies that effectively address the unique dietary challenges and health disparities faced by diverse populations worldwide [11, 21–23].

This chapter aims to address the challenges posed by the diverse approaches and methods used in collecting dietary data at local, national, and global levels. While numerous programs and dedicated researchers are actively engaged in estimating global dietary intakes [24–27], the lack of uniformity in data collection methods hinders comprehensive and standardized analyses. In response to this, our goal is to systematically introduce methods for comparing dietary intakes both at the individual and population levels. By incorporating social determinants analysis, we aim to identify vulnerable populations, thereby providing deeper insights into global dietary dynamics and health inequities.

To illustrate the individual-level comparison, we focus on sodium intake using data from the National Health and Nutrition Examination Survey (NHANES) for the United States and the China Health and Nutrition Survey (CHNS) for China. High sodium intakes have become one of the major dietary risk factors, as evidence has shown a strong association with chronic diseases such as hypertension, cardiovascular disease, heart attacks, and strokes [28–30]. Although there is growing evidence suggesting that excessive sodium intake is associated with many health issues, and many countries have implemented sodium reduction policies, actual consumption remains high worldwide [30]. In addition, there is still controversy over the recommended amount of salt intake. While the American Heart Association recommends no more than 1.5 g of sodium per day [31], the WHO recommends no more than 2 g of sodium per day [32], and Chronic disease risk reduction intake (CDRR) recommends no more than 2.3 g of sodium per day. A recently published study, however, suggests a different threshold. According to this study, blood pressure begins to increase with sodium consumption of 5 g per day, and it remains stable when intake is below 4 g per day [33].

Shifting from this individual-level analysis to a broader perspective, we leverage the Global Dietary Database (GDD) and the Global Burden of Disease (GBD) for population-level comparisons. Through this comparative approach, we provide valuable insights into the intricacies of dietary patterns, allowing for a more nuanced understanding of global health disparities and paving the way for targeted interventions that address specific needs across diverse populations.

2 Individual Level Assessment and National Dietary Database Comparison: NHANES and CHNS

2.1 The National Health and Nutrition Examination Survey (NHANES)

The National Health and Nutrition Examination Survey (NHANES), conducted by the Centers for Disease Control and Prevention, aims to assess the health and nutritional status of adults and children in the United States [34]. The survey recruits a sample population from all age groups across all 50 states and Washington D.C. Each year, participants from 15 states are visited in-person. NHANES intentionally oversamples individuals who are 60 years and older, African Americans, Asians, and the Hispanic population to ensure statistically reliable data [27, 34]. Dietary data in NHANES are collected using a two-day 24-hours recall method, where Day 1 data are gathered at the Mobile Examination Center and Day 2 data are collected via phone call 3–10 days later with ~87% of participants provided data for both days [27]. In addition to dietary intakes, information on dietary supplements is also collected [34]. Furthermore, each survey cycle includes a Food Frequency Questionnaire (FFQ) with a varying number of questions, focusing on different dietary aspects in each cycle [27].

2.2 China Health and Nutrition Survey (CHNS)

CHNS is a nationwide ongoing open cohort, designed to examine the effects of the health, nutrition, economic, and family planning policies and programs. Initially, up to 2011, the study included samples from nine provinces: Liaoning, Heilongjiang, Jiangsu, Shandong, Henan, Hubei, Hunan, Guangxi, and Guizhou [35]. Since 2011, three megacities—Beijing, Chongqing, and Shanghai—have been added to the cohort, followed by three additional provinces—Shanxi, Yunnan, and Zhejiang—since 2015 [35, 36]. The samples were selected using a multistage, stratified, random cluster method. Although not all provinces in China are represented in the survey, it is still considered nationally representative as it covers states with a significant portion of the Chinese population [36]. Individual dietary intake data were collected over the same three consecutive days for adults aged between 20 and 45 in 1989 and for all individuals in subsequent years [26]. The CHNS employed the 24-hours recall method, gathering data through interviews.

According to the survey reports, over 99% of the sample provided complete data for all three days [26]. This high response rate indicates the reliability and thoroughness of the data collection process, offering a comprehensive insight into the dietary habits of the Chinese population across different periods.

2.3 Comparison of NHANES and CHNS

We provide a brief comparison of these two databases with respect to key characteristics valuable for understanding similarities and differences (Table 1). According to Table 1, dietary intakes across various age and sex groups for the years 2000 (1999–2000), 2004 (2003–2004), 2006 (2005–2006), 2009 (2009–2010), 2011 (2011–2012), and 2015 (2015–2016) are consistent, both in terms of overall consumption and within each subgroup categorized by age (over 2 years old) and sex. This consistency is maintained after adjustments are made for year, units, and weighting. However, it is not feasible to make similar comparisons based on race groups between the US and China due to their significant differences. In terms of educational groups, despite some differences in the educational systems of the US and China, it is possible to align corresponding educational levels. For instance, 'High school graduate/GED or equivalent' in the US is comparable to 'Upper middle school degree' in China, and 'College graduate or above' in the US is equivalent to 'University or college degree and master's degree or higher' in China. Comparing annual household incomes is problematic due to variations in units, living costs, and lifestyles. In terms of dietary factors, while the CHNS reports overall intakes for energy and three macronutrients, it does not include micronutrients. Nonetheless, data from 24-hours recalls and Food Frequency Questionnaires (FFQ) provided by CHNS can be used to estimate vitamin and mineral intake.

2.4 Illustrative Example: Sodium Intake Among Chinese and American Adults

To demonstrate the challenges and opportunities in comparing individual intakes, this analysis utilizes NHANES and CHNS as case studies, specifically focusing on sodium intake in the United States and China. By examining these country-representative datasets from the two countries, we aim to understand the methodological approaches and assess the public health implications, particularly in terms of health equity. Additionally, this analysis highlights the complexities involved in conducting cross-country comparisons.

Table 1 Comparison of basic characteristics in the two individual-level databases, NHANES and CHNS

	The National Health and Nutrition Examination Survey (NHANES) [27, 34]	China Health and Nutrition Survey (CHNS) [26, 35, 36]
Representative country	United States, covering all 50 states and Washington D.C	China, covering 13 provinces and 3 megacities
Data collection period	11 cycles	10 time points
Cycles	1999–2000, 2001–2002, 2003–2004, 2005–2006, 2007–2008, 2009–2010, 2011–2012, 2013–2014, 2015–2016, 2017–2018, 2019–2020[a]	1989, 1991, 1993, 1997, 2000, 2004, 2006, 2009, 2011, 2015, 2019
Demographics	Age, Sex, Race, Residency, Education level, Annual household income	Age, Sex, Race, Residency, Education level, Annual household income
Age	0–79, and above 80 years old	Over 2 years old
Sex	Male/Female	Male/Female
Race	6 groups	56 groups in total
	Mexican American	
	Other Hispanic	
	Non-Hispanic White	
	Non-Hispanic Black	
	Non-Hispanic Asian	
	Other race—including multi-racial	
Residency	Geographic variables are suppressed	Province data are suppressed, but residency (urban/rural) are released
Educations level	5 groups	6 groups
	Less than 9th grade	Graduated from primary school
	9-11th grade	Lower middle school degree
	High school graduate/GED or equivalent	Upper middle school degree
	Some college or AA degree	Technical or vocational degree, University or college degree
	College graduate or above	Master's degree or higher
Annual household income	14 Groups[b]:	Report using 0–9,999,999 yuan

(continued)

Table 1 (continued)

	The National Health and Nutrition Examination Survey (NHANES) [27, 34]	China Health and Nutrition Survey (CHNS) [26, 35, 36]
	$ 0 to $ 4999	
	$ 5000 to $ 9999	
	$10,000 to $14,999	
	$15,000 to $19,999	
	$20,000 to $24,999	
	$25,000 to $34,999	
	$35,000 to $44,999	
	$45,000 to $54,999	
	$55,000 to $64,999	
	$65,000 to $74,999	
	$20,000 and over	
	Under $20,000	
	$75,000 to $99,999	
	$100,000 and over	
Dietary assessment methods	2-day 24HR recall + FFQ	3-day 24HR recall + FFQ
Dietary factors	Energy (kcal)	Energy (kcal)
	Protein (g)	Protein (g)
	Carbohydrate (g)	Carbohydrate (g)
	Total sugars (g)	Total fat (g)
	Dietary fiber (g)	Other nutrients[c]
	Total fat (g)	
	Total saturated fatty acids (g)	
	Total monounsaturated fatty acids (g)	
	Total polyunsaturated fatty acids (g)	
	Cholesterol (mg)	
	All Vitamins and Minerals (mg)	
	Total plain water (g)	
	Total tap water (g)	
	Total bottled water (g)	

[a] Data collection for the 2019–2020 cycle is incomplete due to the COVID-19 pandemic [37]
[b] Classifications of income vary by cycle; the classification shown in this table is from the 2017–2018 cycle
[c] Intakes of other nutrients are not provided by CHNS but can be retrieved from 24HR recall data

2.5 Sodium Intake Among Chinese Adults

For the Chinese population, our analysis is based on sodium intake calculations by Yu et al., using CHNS data from the years 1997, 2000, 2004, 2006, 2009, and 2011 [38]. The daily consumption of household salt was determined by tracking changes in the home's food stock [35, 38]. Throughout each survey period, any purchases of salt from markets were accurately recorded and measured with digital scales accurate to 0.1 g, from the beginning to the end of the survey [35, 38]. This included taking into account the initial stock of salt at the beginning of the survey, purchases made during the survey, and any losses during food preparation, all of which contributed to the final calculation of household consumption [38]. The distribution of salt among household members was based on their respective Recommended Nutrient Intakes (RNIs) [38]. They calculated each consumption by dividing their RNI by the total RNI for the household [38]. This proportion was then used to allocate the consumption of salt to each household member [38].

They found out that the consumption of salt in China has dropped significantly from 1997 to 2011 [38]. Based on the results they provided (Table 2 displayed in this study), the average daily salt intake notably declined from 12.9 ± 0.1 g/day to 8.4 ± 0.1 g/day ($\Delta = 4.5$ g/day) [38]. Specifically, the intake among males fell from 13.9 ± 0.1 g/day to 9.0 ± 0.1 g/day ($\Delta = 4.9$ g/day), while female intake dropped from 12.0 ± 0.1 g/day to 7.9 ± 0.1 g/day ($\Delta = 4.1$ g/day) [38]. Additionally, the study provided a detailed analysis of salt intake among females and males, further categorized by two age groups: 18–44 years and 45–59 years [38].

In addition to salt intake, we learn about the changes in sodium intake for both gender groups and two age groups. Specifically, a significant reduction in overall sodium intake, decreasing from 5.3 ± 0.03 g/day to 3.8 ± 0.02 g/day ($\Delta = 1.5$ g/day) [38]. The sodium intake among males fell from 5.7 ± 0.05 g/day to 4.0 ± 0.03 g/day ($\Delta = 1.7$ g/day), while female intake dropped from 4.9 ± 0.05 g/day to 3.5 ± 0.03 g/day ($\Delta = 1.4$ g/day) [38]. Like salt intake, this study provides average sodium intake for both gender groups and further stratified by two age groups as depicted in Table 2 [38]. However, it does not provide a detailed breakdown of overall sodium intake exclusively stratified by these age groups. Additionally, the researchers offer insights into salt and sodium intake differentiated by activity level and region (urban/rural).

This analysis also provides an estimated proportion of participants with excessive salt consumption, following the criteria set by the WHO criteria (<5 g salt/day) and the Chinese DRI (<6 g salt/day) [38–40]. Throughout the six survey periods, more than 75% of males were consistently found to consume an excessive amount of salt (Table 2) [38]. In contrast, the proportion of females exceeding the recommended salt intake remained above 70.0% in all cycles, except for 2011, according to the WHO criteria [38]. Notably, when the Chinese DRI standards are applied, there is a discernible decrease in the percentage of the population categorized as having excessive salt intake [38]. This reduction is attributed to the less restrictive criteria set by the Chinese DRI.

Table 2 Salt and Sodium Consumption among the Chinese adults, CHNS 1997–2011 [38]

	1997	2000	2004	2006	2009	2011	Δ[a]
	Mean (SE)	Mean (SE)	Mean (SE)	Mean (SE)	Mean (SE)	Mean (SE)	
Salt intake (g/d)							
Total	12.9 (0.1)	12.2 (0.1)	10.1 (0.1)	9.3 (0.1)	8.9 (0.1)	8.4 (0.1)	4.5
Male							
Overall	13.9 (0.1)	13.1 (0.1)	10.9 (0.1)	9.9 (0.1)	9.6 (0.1)	9.0 (0.1)	4.9
18–44 years	14.1 (0.2)	13.2 (0.2)	10.8 (0.2)	10.2 (0.2)	9.5 (0.1)	9.1 (0.1)	5.0
45–59 years	14.5 (0.3)	13.6 (0.2)	1.7 (0.2)	10.7 (0.2)	10.3 (0.2)	9.4 (0.1)	5.1
Female							
Overall	12.0 (0.1)	11.3 (0.1)	9.4 (0.1)	8.7 (0.1)	8.4 (0.1)	7.9 (0.1)	4.1
18–44 years	12.1 (0.2)	11.3 (0.1)	9.5 (0.1)	9.0 (0.1)	8.3 (0.1)	8.0 (0.1)	4.1
45–59 years	12.2 (0.2)	11.5 (0.2)	9.5 (0.1)	8.7 (0.1)	8.7 (0.2)	8.0 (0.1)	4.2
Sodium intake (g/d)							
Total	5.3 (003)	4.9 (0.03)	4.5 (0.03)	4.1 (0.03)	4.0 (0.03)	3.8 (0.02)	1.5
Male							
Overall	5.7 (0.05)	5.3 (0.04)	4.9 (0.04)	4.4 (0.04)	4.3 (0.04)	4.0 (0.03)	1.7
18–44 years	5.8 (0.07)	5.3 (006)	4.8 (006)	4.5 (0.06)	4.3 (0.06)	4.1 (0.06)	1.7
45–59 years	5.9 (0.1)	5.5 (009)	5.3 (008)	4.7 (0.07)	4.6 (0.08)	4.2 (0.05)	1.7
Female							
Overall	4.9 (0.05)	4.6 (0.04)	4.2 (0.04)	3.8 (0.03)	3.8 (0.04)	3.5 (0.03)	1.4
18–44 years	5.0 (006)	4.6 (005)	4.2 (005)	4.0 (0.05)	3.7 (0.06)	3.6 (0.05)	1.4
45–59 years	5.0 (009)	4.7 (007)	4.3 (0.06)	3.9 (0.06)	3.9 (0.07)	3.5 (005)	1.5
Proportion (WHO: < 5 g salt/day)[b]							
Male	87.8%	91.4%	84.0%	81.7%	79.7%	75.8%	
Female	84.2%	87.4%	78.5%	76.2%	72.9%	69.9%	
Proportion (Chinese DRI: < 6 g salt/day)[b]							
Male	82.7%	85.8%	75.8%	71.9%	69.3%	65.7%	
Female	77.8%	79.8%	67.7%	64.3%	59.9%	58.1%	

[a] Change between 1997 and 2011
[b] Proportion of people consuming an excessive amount of salt intake

Several other studies have also reported on salt intake among Chinese adults. For instance, Zhang et al. documented salt intake in the years 1991, 2000, 2009, and 2018, utilizing the CHNS database and calculating salt intake using a procedure similar to that of Yu et al. [38, 41]. The primary difference in their methodology was the allocation of household salt proportions to individuals based on the proportion of total energy consumed by each person, rather than using the DRI [41]. However, their reporting approach differs slightly. Instead of presenting the mean intake, they reported the median, 25th percentile (p25), and 75th percentile (p75) for overall salt intake, as well as intakes stratified by age (18–44, 45–59, and ≥ 60 years old), gender (male/female), area (urban/rural), and region (north/south) [41]. While the same database and a similar method were used, the results are only partially comparable to those of Yu et al.'s study. Only the overall intakes and those stratified by gender and area (urban/rural) for the years 2000 and 2009 are directly comparable. For example, Zhang et al. reported an overall daily median intake in 2000 and 2009 of 10.0 g/day (p25: 6.8, p75:14.8) and 7.1 g/day (p25:4.8, p75:10.2), respectively [41]. In contrast, Yu et al. reported overall daily mean intakes of 12.2 ± 0.1 g/day and 8.9 ± 0.1 g/day [38]. Although both studies indicate a decreasing trend, the reported results differ by 18.0 and 20.2%.

This discrepancy underscores the importance of reporting full descriptive statistics, including minimum, 25th percentile, median, mean, standard deviation (SD), 75th percentile, and maximum values of intakes, rather than just the mean value. Providing comprehensive descriptive statistics allows for a deeper understanding of the population distribution, which is particularly vital considering that the distribution curve of dietary intakes tends to be right-skewed [42], but also enables the identification of outliers and the assessment of data variability.

2.6 Sodium Intake of the US Adults

To facilitate a comparison with US sodium intake (in grams), an ideal approach would be to use a table with a layout identical to or similar to that used in the study by Yu. However, despite numerous studies utilizing data from the National Health and Nutrition Examination Survey (NHANES), there appears to be limited study that presents data in a format like that of Yu's study. For examining the US population, we refer to the study conducted by Hu et al., which focuses on sodium intake among US adults using NHANES data from 2005–2006 and 2015–2016 [43]. As previously mentioned, NHANES reports mineral intakes for each participant based on their 24-hour recalls, with researchers estimating the daily mean intake using the average from the two-day 24-hour recalls in each survey period [34]. Additionally, this study examines the prevalence of individuals exceeding the recommended sodium thresholds in the total US population. The threshold used is less than 2300 mg of sodium per day, as per the 2015–2020 Dietary Guidelines for Americans (DGA), equivalent to 5.85 g of salt (1 g sodium * 2.542 = 1 g salt) [32, 44].

This study reports the overall daily mean sodium intake, stratified by age (18–29, 30–39, 40–49, 50–59, 60–69, 70–79, ≥ 80 years old), sex (male/female), race, and energy intake (over 2100 kcal, and under 2100 kcal) for both survey periods [43]. In 2005–2006, the overall sodium intake was 3479.3 mg/day (95% CI: 3408.9, 3549.6) [43]. Among male adults, the intake was 4066.2 mg/day (95% CI: 3958.0, 4174.4), and for female adults, it was 2921.3 mg/day (95% CI: 2865.3, 2977.3) [43]. Furthermore, it was observed that 76.9% of US adults (95% CI: 75.1%, 78.5%) did not meet the recommended sodium intake thresholds, with

86.8% of males (95% CI: 84.9%, 88.5%) and 67.8% of females (95% CI: 65.4%, 70.1%) falling into this category [43]. In comparison, the overall sodium intake in 2015–2016 was 3514.2 mg/day (95% CI: 3431.2, 3597.2), with male adults consuming 4061.8 mg/day (95% CI: 3937.5, 4186.1) and female adults consuming 2995.2 mg/day (95% CI: 2922.9, 3067.4) [43]. In this period, 79.1% of US adults (95% CI: 76.8%, 88.5%) did not meet the recommended thresholds, including 88.5% of males (95% CI: 86.3%, 90.4%) and 70.3% of females (95% CI: 67.0%, 73.5%) [43]. While there was a decrease in absolute sodium intake in 2005–2006, the percentage of the population not meeting the DGA increased. This discrepancy highlights the vital importance of thoroughly understanding the distribution of sodium intakes across the population.

2.7 Comparison of Intakes in Two Countries Using Individual-Level Databases: Strengths and Limitations

The comparable year between Yu et al.'s study and Hu et al.'s study is CHNS 2006 versus NHANES 2005–2006 [38, 43]. However, due to the differing age groups reported in these studies, it is not feasible to compare stratified intakes by age. On the other hand, stratified intakes by sex are comparable. In CHNS 2006, the overall intake was 4.1 ± 0.03 g/day, with average daily sodium intake among Chinese males at 4.4 ± 0.04 g/day and females at 3.8 ± 0.03 g/day [38]. In contrast, NHANES 2005–2006 reported lower sodium intakes for the US population, 3.48 g/day (95% CI: 3.41, 3.55), with male adults consuming an average of 4.07 g/day (95% CI: 3.96, 4.17) and female adults 2.92 g/day (95% CI: 2.87, 2.98) [43]. The overall difference is 7.3%, 7.5% for males, and 23.1% for females. During this period, the proportion of people not meeting the recommended thresholds (Chinese DRI: ≤ 2.36 g/day and US DGA: ≤ 2.3 g/day) was 71.9% for Chinese males, 64.3% for Chinese females, 86.8% for US males, and 67.8% for US females [38, 43]. The significant difference in consumption patterns between the two countries, particularly among females, is evident in terms of absolute values. Moreover, the high percentage of individuals exceeding the recommended sodium thresholds in both countries, especially among males, emphasizes the ongoing challenges in dietary management and public health. This comparative analysis highlights the distinct dietary behaviors and health implications in the US and China, underscoring the need for tailored public health strategies in each country.

2.8 Limitations in Existing Individual Level Databases

While we established parallels for a specific year, such as CHNS 2006 compared to NHANES 2005–2006, this level of comparability does not extend to other years. Comparing dietary intakes, sodium intake in this case, between the US and China within a certain time frame, presents significant challenges. These challenges are primarily due to the variations in survey years, age group categorizations, and different definitions of urban/rural, race, and income levels. The methodologies and data collection periods of these studies differ, creating gaps that hinder a comprehensive longitudinal comparison. Such variations are particularly evident in the reporting of age groups, which complicates direct comparisons across similar demographics. Furthermore, while sodium is one of the easier nutrients to compare due to its direct measurability using scales, discrepancies still exist. For instance, China reports salt intake in grams, and the US reports sodium intake in milligrams. Although conversions between these units are possible, potential errors could impact the results. Additionally, the lack of overall nutrient intake data provided by CHNS adds another layer of complexity to these comparisons.

The challenges in aligning fundamental aspects of dietary data collection and reporting are not just a methodological concern; they hold significant implications for understanding global dietary trends and public health. Inaccurate or superficial comparisons may lead to skewed interpretations or the overlooking of critical nuances in dietary habits, which are pivotal for addressing public health challenges. These challenges highlight the critical need for standardized methodologies and reporting practices in nutritional epidemiology. By facilitating more effective cross-country comparisons, such standardization could ultimately inform better public health strategies and nutritional guidelines, contributing to a more accurate understanding of global dietary trends and their health implications.

3 Population Level Assessment Based on Global Dietary Databases Comparison: GDD Versus GBD

3.1 The Global Dietary Database (GDD)

The Global Dietary Database, constructed by Tufts University, is a comprehensive database that aims to understand global dietary habits, trends, patterns, and diet-related illnesses. It utilizes public, individual-level dietary surveys, such as 24-hour recalls, food frequency questionnaires, and food records, to build a model for estimating the final intakes of various dietary factors. According to GDD, this final model incorporates data from a total of 1225 surveys, representing 185 countries and 99% of the global population. Detailed information about the GDD estimation model is available in the supplemental materials of the paper published by Lara-Castor et al. [24, 45–47].

3.2 The Global Burden of Disease (GBD)

The Global Burden of Disease (GBD) study, led and managed by the Institute for Health Metrics and Evaluation (IHME) at the University of Washington, is a comprehensive and systematic initiative aimed at providing an extensive overview of mortality and disease across various countries, periods, ages, and sexes. This study has become an essential resource for understanding and addressing global health challenges. It identifies fifteen dietary risk factors that significantly increase the risk of diseases. Detailed methodologies and models used in the study, including the approach for assessing these dietary risks, are available in the supplemental materials of the paper published by Stanaway et al. The GBD's insights are invaluable for policymakers, health professionals, and researchers, guiding them in prioritizing health issues, effectively allocating resources, and measuring the impact of various health interventions [25, 45].

3.3 Comparison of GDD and GBD

To facilitate the comparison of country-level dietary intakes, we undertook a thorough process of data harmonization and alignment, focusing on selected food groups, population age, intake units, survey years, and country-specific data. Initially, we conducted a detailed comparison of food group definitions in both GDD and GBD. This led to the identification of thirteen comparable food groups: fruits, vegetables, red meat, processed meat, whole grains, legumes and beans, nuts and seeds, dairy, sugar-sweetened beverages (SSBs), fiber, polyunsaturated fatty acids (PUFAs), calcium, and sodium. We then generated representative values for each food group for adults aged 25 and above. This was achieved by reweighting data from both databases based on the population figures from the United Nations [48].

As Table 3 displays, while dietary intakes stratified by sex are comparable, variables such as education level and urbanity, which are unavailable in the GBD, cannot be compared. Regarding the survey years, we identified four comparable years: 1990, 2000, 2010, and 2017 (with 2018 data from GDD). Moreover, after comparing the list of countries included in both databases, we identified 169 countries for our analysis. In this chapter, we will specifically focus on the case study of sodium intake among adults aged 25 years and older at the country level. The basic characteristics of GDD and GBD are displayed in Table 3.

3.4 Illustrative Example: Sodium Intakes in GDD Versus GBD

As previously mentioned, providing a full range of descriptive statistics is essential for depicting the distribution of dietary intakes, including minimum, 25th percentile, median, mean, standard deviation (SD), 75th percentile, and maximum. Table 4 and Fig. 1 demonstrate that the global mean intake of sodium across the combined four years is 2.68 ± 0.74 g/day according to GDD. This is supported by a minimum intake of 0.95 g/day, a median of 2.63 g/day, and a maximum intake of 7.5 g/day. Further insights into the distribution are provided by the skewness and kurtosis values, which are 0.13 and 0.21, respectively. In contrast, data from GBD indicated a global mean intake of 3.88 ± 1.15 g/day, with a minimum intake of 1.99 g/day, a median of 3.76 g/day, and a maximum intake of 10.67 g/day. The skewness and kurtosis are 0.22 and 0.19, respectively. Reporting only the mean

Table 3 Comparison of basic characteristics in the two global dietary databases, GDD and GBD

Database	Global Dietary Database (GDD) [24, 45, 46]	Global Burden Disease (GBD) [25, 45]
Food group	54 dietary factors	15 dietary risk factors
Level	Global, Geographic region, Country	Global, Country
Demographic granularity	Age, Sex, Education, Urbanicity	Age, Sex
Age	All ages	25–29 years,
	0–11 month, 12–23 month, 2–5 years	30–34 years, etc
	6–10 years, 11–14 years, 15–19 years	Every 5-year interval for adults until 95 +
	20–24 years, etc. and every 5-year interval for adults until 95 +	
Sex	Male/Female	Male/ Female
Education	Low (0–6 years formal)	Not available
	Medium (6.01–12 years)	
	High (12.01 + years)	
	All education levels	
Urbanicity	Rural	Not available
	Urban	
	All residences	
Year	1990, 1995, 2000, 2005, 2010, 2015, 2018	1990, 2000, 2010, 2017
Data source	National nutrition surveys: dietary recalls and records, food-frequency questionnaires	Nationally representative nutrition surveys; national sales data
Model used	Bayesian multilevel framework	Bayesian meta-regression model
Energy adjustment	0–0.9 years, 700 kcal	25 + years, 2000 kcal
	1–1.9 years, 1000 kcal	
	2–5 years, 1300 kcal	
	6–10 years, 1700 kcal	
	11–74 years, 2000 kcal	
	75 + years, 1700 kcal	

value without the range can obscure the full picture of dietary sodium intake distribution, potentially leading to misinterpretations or oversimplified conclusions about the true nature of dietary habits and risks.

We systematically compared sodium intakes across 169 countries for the years 1990, 2000, 2010, and 2018 [49] (Tables 4 and 5). This comparison used

Table 4 Descriptive statistics of global, US, and China sodium intakes in both the Global Dietary Database (GDD) and Global Burden of Disease (GBD) for the combined years, 1990, 2000, 2010, and 2018

Year	Database	Mean (SD)[a]	% Δ[b] (%)	Spearman test		Min	P25[d]	Median	P75[e]	Max
				R[c]	P-value					
All	GDD	2.68 (0.74)	−36	0.56	<0.001	0.95	2.23	2.63	3.04	7.50
	GBD	3.88 (1.15)				1.99	3.05	3.76	4.38	10.67
1990	GDD	2.67 (0.72)	−41	0.48	<0.001	1.07	2.23	2.65	3.00	7.39
	GBD	4.06 (1.39)				2.04	3.14	3.70	4.40	10.51
2000	GDD	2.66 (0.74)	−37	0.53	<0.001	0.95	2.24	2.61	2.99	7.41
	GBD	3.88 (1.19)				1.99	3.10	3.69	4.42	10.23
2010	GDD	2.69 (0.76)	−35	0.60	<0.001	0.97	2.28	2.63	3.06	7.50
	GBD	3.81 (1.00)				2.05	2.93	3.82	4.36	10.53
2018	GDD	2.70 (0.75)	−33	0.63	<0.001	0.96	2.20	2.66	3.07	7.06
	GBD	3.77 (0.96)				2.16	2.92	3.82	4.36	10.67
USA	GDD	3.26 (0.01)	−37	0.2	0.8	3.25	3.25	3.26	3.27	3.28
	GBD	4.72 (0.53)				3.99	4.52	4.90	5.10	5.11
China	GDD	4.29 (0.26)	−84	0.6	0.4	3.91	4.20	4.37	4.46	4.49
	GBD	10.49 (0.18)				10.23	10.44	10.52	10.57	10.67

Notes Units are in grams/day

[a] *SD* standard deviation

[b] % Δ—percent difference = (mean $_{GDD}$ − mean $_{GBD}$)/((mean $_{GDD}$ + mean $_{GBD}$)/2) * 100%

[c] *R* Spearman correlation coefficient

[d] P25—25th percentile

[e] P75—75th percentile

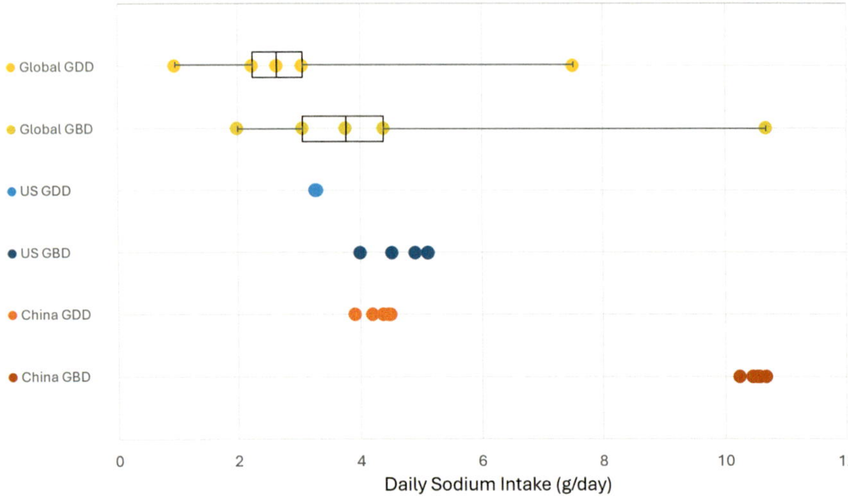

Fig. 1 Descriptive statistics for global, US, and China sodium intakes in both Global Dietary Database (GDD) and Global Burden of Disease (GBD) for the combined years 1990, 2000, 2010, and 2018. For global intakes, box-plots are marked with five points representing the minimum, 25th percentile, median, 75th percentile, and maximum, from left to right, respectively. For comparison, the estimated minimum, 25th percentile, 75th percentile, and maximum (from left to right) intakes in US and China, are shown for GDD and GBD

distribution-based descriptive statistics, the Bland–Altman method, and an evaluation against the recommended sodium intake threshold of 2.3 g per day [44, 49, 50]. Our analysis led to the conclusion that over 80% of the global population consumes sodium at levels exceeding the recommended threshold [49]. Additionally, we identified some inconsistencies between the two databases, highlighting areas that may require further investigation and alignment in future dietary data compilations [50–52].

3.5 Limitations in Population Level Existing Databases

Despite the extensive data and insights provided by GDD and GBD, there are inherent limitations in these population-level databases that pose challenges for global dietary analysis. One of the primary issues is the variability in methodologies and data collection approaches between these two databases. They used different original data resources which used various methods to collect the dietary intake data. This variability can lead to inconsistencies in the comparison of dietary data, such as differences in the categorization of food groups, variations in dietary assessment methods, and discrepancies in the definition and classification of dietary factors and demographic variables like age, sex, and socioeconomic status. Additionally, while both databases provide extensive global coverage, data gaps for certain regions and time periods can limit the scope of longitudinal and

Table 5 Descriptive statistics of global, US, and China sodium intakes in both the Global Dietary Database (GDD) and Global Burden of Disease (GBD) for individual years, 1990, 2000, 2010, and 2018

Year	Level	Database	Mean (SD) [a]	% D[b]
1990	Global	GDD	2.67 (0.72)	−41
		GBD	4.06 (1.39)	
	US	GDD	3.25	−20
		GBD	3.99	
	China	GDD	3.91	−91
		GBD	10.51	
2000	Global	GDD	2.66 (0.74)	−37
		GBD	3.88 (1.19)	
	US	GDD	3.28	−35
		GBD	4.69	
	China	GDD	4.30	−82
		GBD	10.23	
2010	Global	GDD	2.69 (0.76)	−35
		GBD	3.81 (1.00)	
	US	GDD	3.26	−44
		GBD	5.10	
	China	GDD	4.49	−80
		GBD	10.53	
2018	Global	GDD	2.70 (0.75)	−33
		GBD	3.77 (0.96)	
	US	GDD	3.25	−44
		GBD	5.11	
	China	GDD	4.45	−82
		GBD	10.67	

geographical comparisons. Furthermore, although descriptions of models and final estimates are provided for both databases, access to the simulation data is limited, and the entire estimation process is not entirely replicable. These limitations are not only technical but have broader implications for understanding global dietary patterns and their impact on public health. Inaccuracies or inconsistencies in data collection and analysis can lead to misinformed conclusions. Therefore, there is a pressing need for improved standardization and harmonization of data sources and methodologies across these databases. Such advancements would facilitate more accurate and comprehensive comparisons of dietary data on a global scale. At the very least, providing detailed methodology and accessible simulation data would enhance the transparency of the data, contributing to more reliable and informative dietary health assessments.

4 Disadvantaged Population, Health Equity, and Nutrition Surveillance

In analyzing dietary data from sources like the CHNS and NHANES, a critical limitation becomes apparent: the underrepresentation of disadvantaged populations. These surveys primarily focus on community-dwelling individuals, thereby excluding groups that are not easily accessible, including the homeless, hospitalized individuals, and those residing in nursing homes [53]. This oversight is concerning as it leads to a lack of data on some of the most vulnerable and disadvantaged groups. These populations often face substantial barriers to accessing healthy food choices and encounter significant challenges in managing dietary risks, such as reducing sodium intake. The absence of their representation in these surveys suggests a potential gap in our understanding of the full spectrum of dietary habits and health risks, especially among those who are most in need of nutritional interventions and support.

Furthermore, while the GDD and GBD cover an impressive range of 184 countries, disparities within and across these countries can be substantial. Relying on a single database for dietary analysis may lead to erroneous decisions, particularly in identifying the most vulnerable populations who require more targeted interventions and policy changes. This is exemplified in Fig. 2, which shows the top five countries with the largest discrepancies and their comparison to the recommended intakes. Especially for Indonesia, Mauritius, and Viet Nam, GDD concludes that these countries' average dietary intakes fall within recommended thresholds, while GBD draws the opposite conclusion. Such discrepancies can misguide the direction of policy implementation and intervention programs.

Furthermore, it is crucial to recognize that several countries are not represented in these two databases (Fig. 3), leaving a gap in our understanding of dietary patterns and health equity in these regions. Addressing these gaps is essential not only for comprehensive data collection but also for the development of effective

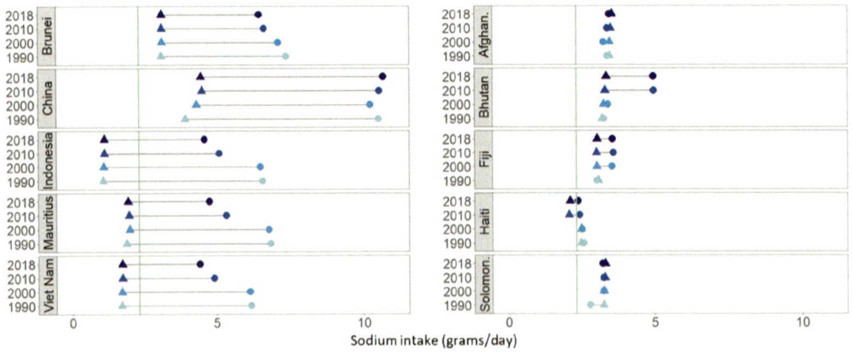

Fig. 2 The top 5 countries with the largest (left) and smallest (right) discrepancies of GDD estimations and GBD estimations in sodium consumption

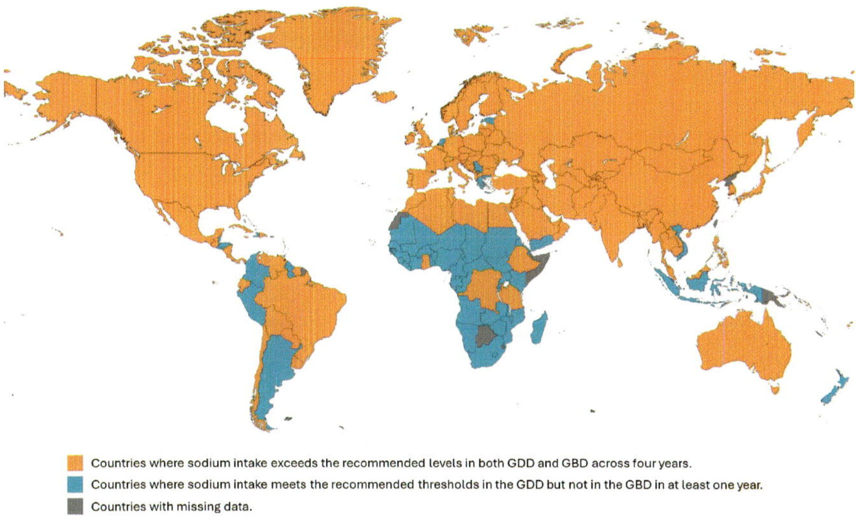

Fig. 3 Map illustrating the consistency of estimated sodium intakes at the country level between the global dietary database and the global burden of diseases, with the absence of data for certain countries/regions. The orange color indicates countries where sodium intake exceeds the recommended levels in both databases across four years. The blue color denotes countries where sodium intake meets the recommended thresholds in the GDD but not in the GBD for at least one year. Grey color marks countries with missing data. Notably, there are no countries where sodium intake meets the thresholds in the GBD but not in the GDD

public health strategies and the advancement of health equity on a global scale. Additionally, the countries highlighted in orange in Fig. 3 represent those with excessive intake in both databases, while those in blue indicate countries with excessive intake in at least one database. The recommended threshold we used is 2.3 g of sodium/day [44]. Besides highlighting the inconsistencies across various countries, it is also important to consider the proportion of the affected population in these countries [49]. A country with a large population does not represent a singular entity but rather a significant number of individuals impacted by these dietary trends.

5 Conclusion and Moving Forward

Comparing sodium intake between the US and China using individual-level databases presents significant challenges, such as aligning data across different years, units, demographic variables, and dietary assessment methods. This difficulty is compounded at the population level, where these databases show notable discrepancies in country-level sodium intake estimations. Moreover, the common practice of releasing only median intake values, rather than providing a full range of descriptive statistics, limits the depth of analysis possible for sodium intakes.

Furthermore, ongoing debates over the recommended salt intake thresholds highlight the urgent need for more robust, evidence-based dietary guidelines. These challenges collectively emphasize the critical importance of having accurate and comprehensive data in nutrition surveillance systems. High-quality data is essential for effective policy-making and public health interventions, and it is crucial in enhancing our understanding of the nutritional impact on health. In the global pursuit of improved public health outcomes, the precision and reliability of dietary data become essential. Therefore, fortifying nutrition surveillance systems with high-quality data is imperative for making informed decisions and developing strategies that more effectively address health issues on both national and global scales.

From a public health perspective, several types of knowledge gaps are particularly critical, especially those that pertain to data quality and relevance. These gaps include incomplete, inconclusive, inconsistent, and outdated data [54]. An effective nutrition surveillance system must overcome these issues to be truly beneficial, offering data that is representative, current, and accurate. High-quality nutrition surveillance data is indispensable for supporting the development of national policies and the planning of local programs. The development of evidence-based metrics for evaluating databases and data dashboards, along with standardized methodologies, is essential for a comprehensive global understanding of dietary patterns [55, 56]. Rapid development of nutrition science underscores the importance of digital platforms, such as the Global Nutrition and Health Atlas (GNHA), which facilitate large-scale data management and enhance collaborative research [53]. In addition, the expansion of food and nutrition information demands innovative data sharing and visualization methods, as outlined in a systematic review of nutrition interactive dashboards [55]. Without addressing these gaps, nutrition surveillance systems risk falling short of delivering the necessary insights for informed public health decision-making [57]. This shortfall could lead to health interventions that are ineffective or misaligned, potentially exacerbating health inequities instead of mitigating them, and failing to meet the needs of diverse populations.

Artificial Intelligence (AI) has shown promising applications in patient-centered healthcare, with a scoping review revealing numerous studies reporting positive patient perceptions and many of them are used to monitor nutrition status [58]. We might need also to learn how to use novel information technologies to assess food and nutrition-focused policies [59] and to train generative AI to detect suboptimal intakes. These advancements indicate the feasibility of AI's broader application, such as monitoring nutrient intakes for the general public. Such an expansion could revolutionize the way dietary data is collected and analyzed, offering more precise and personalized insights into dietary patterns and nutritional habits.

References

1. Townsend JR, Kirby TO, Marshall TM, Church DD, Jajtner AR, Esposito R (2023) Foundational nutrition: implications for human health. Nutrients 15(13):2837. https://www.mdpi.com/2072-6643/15/13/2837
2. Bailey RL (2021) Overview of dietary assessment methods for measuring intakes of foods, beverages, and dietary supplements in research studies. Curr Opin Biotechnol 70:91–96. https://doi.org/10.1016/j.copbio.2021.02.007
3. Gibney MJ, Margetts BM, Kearney JM, Arab L (2013) Public health nutrition. Wiley
4. Mahaffey KR (1995) Nutrition and lead: strategies for public health. Environ Health Perspect 103(suppl 6):191–196
5. Sahyoun NR (2005) Nutritional surveillanceldeveloped countries. In: Caballero B (ed) Encyclopedia of human nutrition (2nd edn). Elsevier, pp 363–371. https://doi.org/10.1016/B0-12-226694-3/02207-9
6. Brown AGM, Esposito LE, Fisher RA, Nicastro HL, Tabor DC, Walker JR (2019) Food insecurity and obesity: research gaps, opportunities, and challenges. Transl Behav Med 9(5):980–987. https://doi.org/10.1093/tbm/ibz117
7. Micha R, Shulkin ML, Peñalvo JL, Khatibzadeh S, Singh GM, Rao M, Fahimi S, Powles J, Mozaffarian D (2017) Etiologic effects and optimal intakes of foods and nutrients for risk of cardiovascular diseases and diabetes: systematic reviews and meta-analyses from the nutrition and chronic diseases expert group (NutriCoDE). PLoS ONE 12(4):e0175149. https://doi.org/10.1371/journal.pone.0175149
8. Satia JA (2009) Diet-related disparities: understanding the problem and accelerating solutions. J Am Diet Assoc 109(4):610–615. https://doi.org/10.1016/j.jada.2008.12.019
9. Getz GS, Reardon CA (2007) Nutrition and cardiovascular disease. Arterioscler Thromb Vasc Biol 27(12):2499–2506
10. Ross AC, Caballero B, Cousins RJ, Tucker KL (2020) Modern nutrition in health and disease. Jones Bartlett Learn
11. WHO J, Consultation FE (2003) Diet, nutrition and the prevention of chronic diseases. World Health Organ Tech Rep Ser 916(i–viii):1–149
12. Barros FC, Victora CG, Scherpbier R, Gwatkin D (2010) Socioeconomic inequities in the health and nutrition of children in low/middle income countries. Rev Saude Publica 44:1–16
13. Barros FC, Victora CG, Scherpbier RW, Gwatkin D (2010) Health and nutrition of children: equity and social determinants. Equity, Soc Determinants Public Health Programmes 49
14. Ngom P, Binka FN, Phillips JF, Pence B, Macleod B (2001) Demographic surveillance and health equity in sub-Saharan Africa. Health Policy Plan 16(4):337–344
15. Singh GK, Daus GP, Allender M, Ramey CT, Martin EK, Perry C, Andrew A, Vedamuthu IP (2017) Social determinants of health in the United States: addressing major health inequality trends for the nation, 1935–2016. Int J MCH AIDS 6(2):139
16. Winkled G, Döring A, Keil U, Pietinen P, Arveiler D, Cambou JP, Nuttens C, Richard JL, Evans A, McClean R, McCrum E (1992) Comparison of dietary intakes in four selected European populations. Clin Investigator 70(10):889–895. https://doi.org/10.1007/BF00180434
17. Andreeva VA, Deschamps V, Salanave B, Castetbon K, Verdot C, Kesse-Guyot E, Hercberg S (2016) Comparison of dietary intakes between a large online cohort study (Etude NutriNet-Santé) and a nationally representative cross-sectional study (Etude Nationale Nutrition Santé) in France: addressing the issue of generalizability in e-epidemiology. Am J Epidemiol 184(9):660–669. https://doi.org/10.1093/aje/kww016
18. Simpson RB, Gottlieb J, Zhou B, Hartwick MA, Naumova EN (2021) Completeness of open access FluNet influenza surveillance data for Pan-America in 2005–2019. Sci Rep 11(1):795
19. Zhang Y, Simpson RB, Sallade LE, Sanchez E, Monahan KM, Naumova EN (2022) Evaluating completeness of foodborne outbreak reporting in the United States, 1998–2019. Int J Environ Res Public Health 19(5):2898

20. Simpson RB, Gottlieb J, Zhou B, Liang S, Jiang X, Hartwick MA, Naumova EN (2023) Navigating global public influenza surveillance systems for reliable forecasting. In: Rezaei N (ed) Integrated science of global epidemics. Springer International Publishing, pp 111–137. https://doi.org/10.1007/978-3-031-17778-1_6
21. Chopra M, Sharkey A, Dalmiya N, Anthony D, Binkin N (2012) Strategies to improve health coverage and narrow the equity gap in child survival, health, and nutrition. The Lancet 380(9850):1331–1340
22. Black MM, Dewey KG (2014) Promoting equity through integrated early child development and nutrition interventions. Ann N Y Acad Sci 1308(1):1–10
23. Blanck HM, Kim SA (2012) Creating supportive nutrition environments for population health impact and health equity: an overview of the nutrition and obesity policy research and evaluation network's efforts. Am J Prev Med 43(3):S85–S90
24. Miller V, Singh GM, Onopa J, Reedy J, Shi P, Zhang J, Tahira A, Shulkin Morris ML, Marsden DP, Kranz S, Stoyell S, Webb P, Micha R, Mozaffarian D (2021) Global dietary database 2017: data availability and gaps on 54 major foods, beverages and nutrients among 5.6 million children and adults from 1220 surveys worldwide. BMJ Global Health 6(2):e003585. https://doi.org/10.1136/bmjgh-2020-003585
25. Global Burden of Disease Collaborative Network. Global Burden of Disease Study 2019 (GBD 2019) Seattle, United States: Institute for Health Metrics and Evaluation (IHME) (2020)
26. Popkin BM, Du S, Zhai F, Zhang B (2010) Cohort profile: The China Health and Nutrition Survey–monitoring and understanding socio-economic and health change in China, 1989–2011. Int J Epidemiol 39(6):1435–1440. https://doi.org/10.1093/ije/dyp322
27. Ahluwalia N, Dwyer J, Terry A, Moshfegh A, Johnson C (2016) Update on NHANES dietary data: focus on collection, release, analytical considerations, and uses to inform public policy. Adv Nutr 7(1):121–134. https://doi.org/10.3945/an.115.009258
28. Jayedi A, Ghomashi F, Zargar MS, Shab-Bidar S (2019) Dietary sodium, sodium-to-potassium ratio, and risk of stroke: a systematic review and nonlinear dose-response meta-analysis. Clinical Nutrition (Edinburgh, Scotland) 38(3):1092–1100. https://doi.org/10.1016/j.clnu.2018.05.017
29. Patel Y, Joseph J (2020) Sodium intake and heart failure. Int J Mol Sci 21(24):9474. https://doi.org/10.3390/ijms21249474
30. Eilat-Adar S, Buch A, Goldsmith R, Endevelt R, Nitsan L, Blaychfeld-Magnazi M (2023) Salt: a narrative review and local policy initiatives in Israel. J Public Health Policy. https://doi.org/10.1057/s41271-023-00457-y
31. Lloyd-Jones DM, Hong Y, Labarthe D, Mozaffarian D, Appel LJ, Horn LV, Greenlund K, Daniels S, Nichol G, Tomaselli GF, Arnett DK, Fonarow GC, Ho PM, Lauer MS, Masoudi FA, Robertson RM, Roger V, Schwamm LH, Sorlie P, Yancy CW, Rosamond WD (2010) Defining and setting national goals for cardiovascular health promotion and disease reduction. Circulation 121(4):586–613. https://doi.org/10.1161/CIRCULATIONAHA.109.192703
32. World Health Organization (2023) Sodium reduction. World Health Organization. https://www.who.int/news-room/fact-sheets/detail/salt-reduction
33. Shen C, Bowers PJ, Bar-Yam Y (2020) How much sodium should we eat? Prog Prevent Med 5(1):e0026. https://doi.org/10.1097/pp9.0000000000000026
34. Centers for Disease Control and Prevention (CDC). National Center for Health Statistics (NCHS). (2023) NHANES—about the National Health and Nutrition Examination Survey. Centers for Disease Control and Prevention. https://www.cdc.gov/nchs/nhanes/about_nhanes.htm
35. Liang Y, Liu R, Du S, Qiu C (2014) Trends in incidence of hypertension in Chinese adults, 1991–2009: The China Health and Nutrition survey. Int J Cardiol 175(1):96–101. https://doi.org/10.1016/j.ijcard.2014.04.258
36. Chen Y, Kang M, Kim H, Xu W, Lee JE (2023) Associations of dietary patterns with obesity and weight change for adults aged 18–65 years: evidence from the China Health and Nutrition Survey (CHNS). PLoS ONE 18(1):e0279625–e0279625. https://doi.org/10.1371/journal.pone.0279625

37. Centers for Disease Control and Prevention (n.d.) NHANES 2019–2020 overview. Centers for disease control and prevention. https://wwwn.cdc.gov/nchs/nhanes/continuousnhanes/overview.aspx?BeginYear=2019
38. Yu L, Li S, Zhao J, Zhang J, Wang L, Wang K (2018) Secular trends in salt and soy sauce intake among Chinese adults, 1997–2011. Int J Food Sci Nutr 69(2):215–222. https://doi.org/10.1080/09637486.2017.1344203
39. World Health Organization (2012) Guideline: sodium intake for adults and children. World Health Organization. https://www.who.int/publications-detail-redirect/9789241504836
40. Wang SS, Lay S, Yu HN, Shen SR (2016) Dietary guidelines for Chinese residents (2016): comments and comparisons. J Zhejiang Univ Sci B 17(9):649–656. https://doi.org/10.1631/jzus.B1600341
41. Zhang J, Wang H, Wang Z, Du W, Su C, Huang F, Jia X, Ouyang Y, Li L, Wang Y, Jiang H, Ding G, Zhang B (2020) Trends in adult cooking salt intake—China, 1991–2018. China CDC Wkly 2(7):104–108
42. Ruf A, Neubauer AB, Ebner-Priemer U, Reif A, Matura S (2021) Studying dietary intake in daily life through multilevel two-part modelling: a novel analytical approach and its practical application. Int J Behav Nutr Phys Act 18(1):130. https://doi.org/10.1186/s12966-021-01187-8
43. Hu JR, Sahni S, Mukamal KJ, Millar CL, Wu Y, Appel LJ, Juraschek SP (2020) Dietary sodium intake and sodium density in the United States: estimates from NHANES 2005–2006 and 2015–2016. Am J Hypertens 33(9):825–830. https://doi.org/10.1093/ajh/hpaa104
44. U.S. Department of Agriculture and U.S. Department of Health and Human Services (2020) Dietary guidelines for Americans, 2020–2025, (9th edn)
45. GBD 2017 Risk Factor Collaborators (2018) Global, regional, and national comparative risk assessment of 84 behavioral, environmental and occupational, and metabolic risks or clusters of risks for 195 countries and territories, 1990–2017: a systematic analysis for the Global Burden of Disease Study 2017. Lancet 392(10159):1923–1994. https://doi.org/10.1016/s0140-6736(18)32225-6
46. Lara-Castor L, Micha R, Cudhea F, Miller V, Shi P, Zhang J, Sharib JR, Erndt-Marino J, Cash SB, Mozaffarian D, Bas M, Ali JH, Abumweis S, Krishnan A, Misra P, Hwalla NC, Janakiram C, Liputo NI, Musaiger A, Pourfarzi F, Global Dietary Database (2023) Sugar-sweetened beverage intakes among adults between 1990 and 2018 in 185 countries. Nat Commun 14(1):5957. https://doi.org/10.1038/s41467-023-41269-8
47. Khatibzadeh S, Saheb Kashaf M, Micha R, Fahimi S, Shi P, Elmadfa I, Kalantarian S, Wirojratana P, Ezzati M, Powles J, Mozaffarian D (2016) A global database of food and nutrient consumption. Bull World Health Organ 94(12):931–934. https://doi.org/10.2471/blt.15.156323
48. United Nations (n.d.) World population prospects—population division. United Nations. https://population.un.org/wpp/Download/Standard/MostUsed/
49. Chen Y, Zhou B, Naumova E (2024) How well global dietary intake estimates agree: a case of sodium consumption. J Pub Health Policy. https://doi.org/10.1057/s41271-024-00478-1
50. Chen Y, Zhou B, Naumova E (2023) Discrepancies in intake estimates for select dietary factors in two dietary databases: global burden of disease and global dietary database. Curr Dev Nutr 7:101201. https://doi.org/10.1016/j.cdnut.2023.101201
51. Chen Y, Zhou B, Naumova E (2023) Testing the agreement of the eat-lancet diet and dietary intakes from two databases: global dietary database and global burden of disease database. Curr Dev Nutr 7:100625. https://doi.org/10.1016/j.cdnut.2023.100625
52. Chen Y, Zhou B, Naumova E (2023) Assessing nutrition equity by evaluating dietary deficiencies and excesses: a social determinants analysis. In: Presented at the annual conference consortium of universities for global health (CUGH) “global health without borders: acting for impact”, Los Angeles, CA; Mar 7–19
53. Edlin BR, Eckhardt BJ, Shu MA, Holmberg SD, Swan T (2015) Toward a more accurate estimate of the prevalence of hepatitis C in the United States. Hepatology 62(5):1353–1363. https://doi.org/10.1002/hep.27978

54. Naumova EN (2021) Public health inequalities, structural missingness, and digital revolution: time to question assumptions. J Public Health Policy 42(4):531–535. https://doi.org/10.1057/s41271-021-00312-y
55. Zhou B, Liang S, Monahan KM, Singh GM, Simpson RB, Reedy J, Zhang J, DeVane A, Cruz MS, Marshak A, Mozaffarian D, Wang D, Semenova I, Montoliu I, Prozorovscaia D, Naumova EN (2022) Food and nutrition systems dashboards: a systematic review. Adv Nutr 13(3):748–757. https://doi.org/10.1093/advances/nmac022
56. Zhou B, Liang S, Monahan KM, El-Abbadi N, Cruz MS, Chen Y, DeVane A, Reedy J, Zhang J, Semenova I, Montoliu I, Mozaffarian D, Wang D, Naumova EN (2022) An open-access data platform: Global Nutrition and Health Atlas (GNHA). Curr Dev Nutr 6(4):nzac031. https://doi.org/10.1093/cdn/nzac031
57. Tuffrey V, Hall A (2016) Methods of nutrition surveillance in low-income countries. Emerg Themes Epidemiol 13:4. https://doi.org/10.1186/s12982-016-0045-z
58. Asan O, Choi E, Wang X (2023) Artificial intelligence–based consumer health informatics application: scoping review. J Med Internet Res 25:e47260. https://doi.org/10.2196/47260
59. Naumova EN (2023) Who is responsible for AI-generated public health policies? J Public Health Policy 44(4):517–522. https://doi.org/10.1057/s41271-023-00438-1

Revealing Health Disparities with Visual Analytics in the Digital Era: From Individual Attributes to Global Patterns in Nutrition Surveillance

10

Bingjie Zhou, Yutong Chen, and Elena N. Naumova

Abstract

The surge of digital transformation and the advancement in the computational ability to store, analyze, and visualize intricate nutrition and health data has revolutionized the field of nutrition surveillance, significantly expanding its scope and volume. Nutrition and health data are experiencing exponential growth, encompassing a spectrum from omics read-outs to real-time personal app monitoring and global dietary intake patterns. This data expansion unveils unprecedented potential to comprehend the multifaceted dynamics of health and nutrition outcomes across three dimensions—time, space, and populations. The shift also presents challenges of effectively harnessing complex data from multiple sources to identify health and nutrition disparities and transforming information into actionable insights in a more precise and timely manner. This chapter aims to illustrate how these challenges can be transformed into opportunities through visual analytics—a tool that combines computational analysis techniques with interactive visualizations. We describe the current progress of visual analytics, emphasizing the pivotal role of nutrition and health dashboards in revealing health and nutrition disparities at the individual, community, country, and global levels. Furthermore, the chapter sheds light on existing data gaps, specifically addressing data granularity and aggregating data across time and space dimensions coupled with statistical modeling tools to enhance dashboard

B. Zhou · Y. Chen · E. N. Naumova (✉)
Division of Nutrition Epidemiology and Data Science, Friedman School of Nutrition Science and Policy, Tufts University, Boston, MA, USA
e-mail: Elena.Naumova@tufts.edu

B. Zhou
e-mail: Bingjie.Zhou@tufts.edu

Y. Chen
e-mail: Yutong.Chen@tufts.edu

B. W. K. Son (ed.), *The Landscape of Global Health Inequity*, Integrated Science 22,
https://doi.org/10.1007/978-3-031-60502-4_10

capabilities. Through this comprehensive assessment, we provide a holistic view of the multifaceted challenges and opportunities that arise at the intersection of visual analytics and nutrition surveillance.

Keywords

Nutrition surveillance • Visual analytics • Health disparities • Dashboard capabilities

1 Evolution of Nutrition Surveillance: Tracing Its Development Journey

1.1 The History of Global Nutrition Surveillance

The global nutritional surveillance system was initially established at the first World Food Conference in 1974 by the Food and Agriculture Organization (FAO), the World Health Organization (WHO), and the United Nations International Children's Emergency Fund (UNICEF). The aim was to monitor the food and nutrition conditions of the disadvantaged groups of the population at risk and provide a method of rapid and permanent assessment of all factors that influence food consumption patterns and nutritional status [1, 2].

The initiative of this surveillance system was driven by the recognition of the worldwide inequitable distribution of food resources, inadequate information on food consumption patterns and their consequences for nutrition and health status, and the inadequacy of incorporating food and nutritional aspects in national development plans [1]. The foundational step in this nutrition surveillance journey was marked by the first report on the methodology of nutritional surveillance published by the Joint FAO/UNICEF/WHO Expert Committee in 1976. This report outlined the objectives of nutritional surveillance: (1) to describe the nutrition status of the population, with a specific focus on the vulnerable group (2) to explain the causal link to facilitate the selection of preventive measures (3) to advocate for governmental decisions that address the requirements of normal development and emergencies; (4) to predict the progression of nutritional challenges by analyzing present trends; (5) to monitor and evaluate the effectiveness of nutrition programs [2].

Over the years, the concept of nutrition surveillance has evolved. In 1984, Mason et al. defined nutritional surveillance as "to watch over nutrition, in order to make decisions which lead to improvements in nutrition in populations" [3]. Three years later, in 1987, the Administrative Committee on Coordination Subcommittee on Nutrition published the first report on the world nutrition situation with the main objective of providing an agreed basis for raising awareness of malnutrition, which also marked the initial effort to consolidate data on the economic, food supply, and nutritional status at the national level, aiming to illustrate trends at the global and regional levels over the past 25 years [4]. This nutrition situation report was routinely published in the years 1993, 1994, 2004, and 2009. In 2013, WHO provided a technical guide for the development of a food and nutrition surveillance system [5], setting the stage for more structured monitoring. Later, in 2014, the World Health Assembly (WHA) approved the Global Nutrition Monitoring Framework (GNMF) on Maternal, Infant, and Young Child Nutrition which included the following objectives: (1) to monitor progress toward the achievement of the six global targets (for use at global and national levels); (2) to track implementation of selected programs required to achieve the global targets (for use at the global and national level); and (3) to track implementation of all programs required to achieve national targets (for use at national and at sub-national levels). The framework includes six target indicators and 14 core indicators that track intermediate outcomes, processes, and policy [6]. In 2014, the first Global Nutrition Report (GNR) was published. GNR is an independently produced annual stock-take of the state of the world's nutrition. The report tracks global nutrition targets on maternal, infant, and young child nutrition and on diet-related non-communicable diseases (NCDs) adopted by member states of the WHO as well as governments' delivery against their commitments. It aims to make it easier for governments and other stakeholders to make and deliver on high-impact commitments to end malnutrition in all its forms [7]. Most recently, in the year 2021, the WHO-UNICEF Technical Expert Advisory Group on Nutrition Monitoring developed the first-ever global guidance on the National Nutrition Information Systems (NNIS) [8].

1.2 National Nutrition Surveillance Programs Across Different Regions

Following the establishment of a nutrition surveillance system through collaboration among key international organizations, many countries also started to build their own nutrition surveillance systems, considering their unique situations and capabilities and employing various methodologies.

Only a few countries have established nutrition surveillance systems in the EMRO [5]. In 2023, a review of the nutrition surveillance system was conducted in the EMRO region [9]. It should be noted that only 9 out of the 22 countries that were contacted accepted the interview. The countries that agreed to participate were Kuwait, Saudi Arabia, Oman, Morocco, Sudan, Yemen, Syria, and Palestine. The findings of the review highlighted that the EMRO countries predominately relied on secondary data collected from health facilities. Some countries also supplemented their primary dataset with additional information from existing health surveys conducted at sentinel sites, while others adopted a triangulation approach by incorporating data from sources like communities and schools. Regarding the data collection, anthropometrics, micronutrients, food security, and nutrition transient are the primary components of the data collected in the nutrition surveillance system, however, it was observed that most countries only collected one or two indicators under each component. Specifically, wasting, stunting, overweight and obesity, and screening for anemia were covered by all countries included in the study. Notably, the review did not reveal any systematic dietary assessment mentioned among the countries participating in the study.

In another systematic review published in 2023, nutrition surveillance programs were reviewed in 18 middle- and low-income countries from Southeast Asia (SEARO) and Western Pacific (WPRO) regions [10]. Across these countries, the majority have implemented national programs dedicated to gathering data on nutrition and diet-related outcomes. All 18 countries in SEARO and WPRO regions collect anthropometric data, such as weight and height, with an additional inclusion of waist and hip circumferences in most cases. A few countries extended data collection on food insecurity and nutrition transient. All indicators collected across countries were varied by different population subgroups and intervals of surveys. The quantity and types of indicators differed among distinct population subgroups in various countries, and survey intervals also exhibited variations across nations. Another commonality is that all countries in the study periodically collect dietary intake data, mostly using frequency questionnaires, albeit with variations in questionnaire length and a focus on different aspects. Specific instances include questionnaires tailored to the school-based Student Health Survey and the WHO Stepwise approach to noncommunicable disease risk factors. Out of the 18 countries, 8 exclusively collect data on individual nutrition and diet-outcome indicators through an established national nutrition surveillance program on a continuous scale (Bangladesh, China, Indonesia, Lao People's Democratic Republic, Malaysia, Philippines, Thailand, and Viet Nam), 8 countries conducted large-scale periodic national nutrition surveys (Bhutan, Brunei Darussalam, China, Democratic People's Republic of Korea, India, Sri Lanka, Timor-Leste, Thailand and Viet Nam), and 3 countries developed National micronutrient status surveys (Bangladesh, Nepal and Viet Nam).

A review conducted in 2016 found that in Latin America and the Caribbean (LAC) region, 22 of the 46 LAC countries and territories had at least one nationally representative survey (e.g., The Demographic and Health Survey (DHS), Multiple Indicator Cluster Surveys (MICS), Nutrition and Health Survey (NHS), or National Micronutrient Survey (NMS)). These surveys provided anthropometric data like height for age, weight for age, and weight for height in children under 5 years old, offering insights into undernutrition indicators such as stunting, underweight, and wasting. 17 countries provided information on anemia in children under 5 years old and women of reproductive age; 12 had information on anemia in pregnant women. Information on micronutrient deficiencies was available for 25 of the 46 LAC countries/territories and was mainly limited to (1) anemia in children under 5 years old and in WRA and (2) deficiencies in vitamin A and iodine. Limited national representative dietary intake data were available until March 2014. However, the launch of ELANS, a multicenter cross-sectional nutrition, physical activity, and health survey with a nationally representative sample of urban populations from eight Latin American countries (Argentina, Brazil, Chile, Colombia, Costa Rica, Ecuador, Peru, and Venezuela), have started to provide information on dietary intake of key food groups reported to be related to NCDs risk in Latin American countries [11].

In 2015, all 53 Member States of the WHO European Region adopted the European Food and Nutrition Action Plan (FNAP) for 2015–2020. This plan aimed to enhance food system governance, improve the quality of the population's diet and nutritional status, and ultimately promote better health and well-being. Subsequently, the Global Nutrition Policy Review questionnaire was conducted later in 2017 to evaluate the implementation of the FNAP [12]. Out of all the countries in the European region, 94% responded to the questionnaire. Among the countries that responded, 91% reported having growth monitoring programs. Anthropometric measurements were the most common component of such programs, reported by 46% of the countries. Overweight and obesity were the most monitored indicators (26%), which were usually calculated by taking height and weight measurements to determine the body mass index (BMI). In a review of dietary surveys from 53 European countries [13], it was found that 19 countries, mostly from Central and Eastern Europe, lacked nationally representative surveys. Out of the 33 countries surveyed post-2000, only 22 reported energy and nutrient intakes, with macronutrients being more commonly reported than micronutrients.

In many countries in Africa, similar to low-income countries from other regions, nutrition surveillance information comes from two major national large-scale surveys: The Demographic and Health Survey and Multiple Indicator Cluster Surveys [14]. DHS is a household survey funded by the United States Agency for International Development (USAID) since 1984. MICS is the largest household survey program on children and women's health statuses worldwide, developed by UNICEF, and it was first launched in South Africa in 1994. Beyond these national surveys, the Famine Early Warning Systems Network (FEWS NET) Sentinel Surveillance was established in 1985 by USAID in response to severe famines in East and West Africa. This initiative aimed to address the urgent need for improved and more timely alerts regarding potential food security crises. In order to produce timely, accurate, and evidence-based early warning data and analysis of the causes, severity, and consequences of acute food insecurity, the FEWS NET system collected a variety domain of data from acute food insecurity, climate, conflict, migration, market and trade to agroclimatology data [15]. Moreover, the Standardized Monitoring and Assessment of Relief and Transitions (SMART) survey has been widely conducted in Sub-Saharan Africa regions since 2006, focusing on rapid assessment of nutritional status, especially acute malnutrition in children under five [16]. It is worth noting that, however, limited information on dietary intake data on the individual level was provided in any of those surveillance programs.

North America has a longer and more consistent history of national nutrition surveillance, predominately anchored by The National Health and Nutrition Examination Survey (NHANES). This major national survey has been monitoring the health and nutritional status of adults and children across the United States since its initiation in 1971. NHANES is operated by the National Center for Health Statistics, which is part of the Centers for Disease Control and Prevention (CDC), and has been conducted in two-year cycles since 1999. NHANES collects a wide range of health-related data to assess the health and nutritional status of the

U.S. population, including demographic information, health status and conditions, dietary Intake, anthropometric measurements, and health indicators measured from urine and blood samples [17]. In addition to NHANES, other nutritional surveillance programs in the US have been largely targeted to populations served by maternal and child health programs, with the purpose of watching over trends in undernutrition and iron deficiency [18]. The Pediatric Nutrition Surveillance System (PedNSS) evaluated the prevalence and trends of nutrition-related indicators for low-income children attending federally funded maternal and child health and nutrition programs. Similarly, the Pregnancy Nutrition Surveillance System (PNSS) evaluated the nutritional status of low-income pregnant, postpartum, or nursing women enrolled in federally funded maternal and child health and nutrition programs. Data in these programs are analyzed locally on a monthly basis and/or at the CDC on a quarterly basis [19]. In Mexico, national nutrition and dietary surveys have evolved over the years. The first national nutrition and diet survey (ENN) was conducted in 1988, carried out by the National Institute of Public Health and financed by the Ministry of Health. The second national nutrition survey was done 10 years later. These first two surveys served as a baseline and laid the ground for more recent surveys [20]. Starting in 2006, the National Health and Nutrition Survey (ENSANUT) was formed by combining various health and nutrition surveys. It has been conducted five times so far: in 2006, 2012, 2016, 2018 and 2022. ENSANUT has state, regional, and national representativeness for urban and rural strata for all age groups. Dietary data, anthropometric measurements for the assessment of excess weight for people of all ages, anemia in women and children, and child malnutrition were collected in these surveys [21]. In Canada, the Canadian Community Health Survey (CCHS) initiative began in 2000 and was jointly supported by Health Canada, the Public Health Agency of Canada, Statistics Canada, and the Canadian Institute for Health Information (CIHI). Nutrition was the focused topic in cycles 2004 and 2015, where national and provincial level data on dietary intake, supplement use, anthropometrics, and household food insecurity were collected [22]. 2024 CCHS is currently in progress.

While countries have varying levels of development in their nutrition surveillance systems, there are common key challenges and strategies that span across countries.

- **Sampling Design Strategies Variation**: Nutrition surveillance systems gather information from various data sources. It is important to note that interpreting the data requires caution, and a valid sampling method is crucial to ensure that the data accurately reflects the actual situation within a certain population. However, the variability in sampling design strategies across countries poses a significant challenge to the representativeness of nutrition surveillance data. Secondary data collection, such as health facility-based data, often lacks population sampling and often selected samples. Even though countries conducting surveys employ probability sampling, the methods could still be varied, such as population projections, stratified, and cluster sampling, which may still introduce different sampling errors and biases to the results. The representativeness differences between surveys or programs could pose challenges when comparing or merging nutrition surveillance data obtained through diversity sampling strategies.

- **Spatial Coverage Challenges**: It is not uncommon to see nutrition surveillance systems leave out vulnerable populations residing in certain geographical regions, such as underdeveloped or sparsely populated regions, due to technical difficulties or funding limitations. In addition, limited data was collected among urban deprived areas due to limited health facilities and sentinel sites. This exclusion of certain regions results in underreporting and inequalities in addressing and accessing nutritional needs across varied geographical settings, hindering effective public health policies and targeted interventions to their specific needs, and ultimately preventing the development of a comprehensive nutritional surveillance system that can cater to the needs of the population in diverse geographical locations.
- **Frequency of Data Collection Disparities**: Different time intervals of data collection from various nutrition surveillance programs within or across countries introduce complexities in the timeliness and relevance of nutrition information. Ranging from monthly to annual and 3–5-year intervals, these variations impact the ability to capture dynamic changes in nutritional indicators. To address this issue, a customized method is required, which involves adjusting the frequency of data collection according to the individual features of each indicator. For indicators that change rapidly, like wasting, more frequent surveys within a critical period are crucial. Some indicators affected by seasonal fluctuations need to have a routine weekly or at least monthly to ensure a more precise representation of seasonal patterns.
- **Coverage Disparities in Subpopulations**: There has been a lot of emphasis on monitoring the nutritional status of children under 5 and pregnant women, which is understandable since they are the most vulnerable groups and require the most attention. However, many developed countries have not implemented any significant measures to keep track of the nutritional health of older adults, adult males, adolescents, and school-age children. The definition of "monitoring over time" of nutrition surveillance should go beyond simply monitoring the same indicator over time. It means monitoring individuals over time, which in turn requires us to pay attention to population trajectories. Therefore, it is crucial that we continuously collect and track data for people of all age groups to gain a full understanding of their nutritional status throughout their life cycle. In countries with diverse populations and vastly different socioeconomic statuses, it is essential to collect nutrition surveillance data for different social and economic levels and disaggregate data by subgroups. This is because people in different social and economic groups may be exposed to vastly different food environments and receive different policies associated with their nutritional status. Therefore, subgroup populations may exhibit distinct nutrition patterns that require customized interventions and policies.
- **Advocacy for Multiple Data Sources**: Many studies have been advocating for the integration of multiple data sources from different perspectives. First,

combining secondary data from health facilities and population censuses with primary data from school, community, and national surveys provides a more comprehensive understanding of the nutritional situation. Furthermore, it is imperative for the nutrition surveillance system to broaden its scope beyond gathering information on dietary intake and nutritional status. This expansion should include the examination and data collection on nutrition-related food environments, economic factors, and food systems to provide a holistic understanding of the multifaceted factors contributing to population nutrition and health issues. In addition, it will facilitate a collaborative environment, allowing experts from different domains to work together to address nutrition problems, and it will better help to allocate resources and decrease nutrition and health disparities.

2 Future Direction of Nutrition Surveillance in the Digital Era

The potential for technological advancements in the digital era offers a transformative outlook for nutrition surveillance, significantly expanding its scope and enhancing its capabilities, which paves the way for more targeted and effective policy development based on nutrition surveillance data. This section explores the integration of vast and diverse data sources facilitated by modern technology, offering a new horizon for comprehensively monitoring and understanding nutrition and health at a global scale, and increasing granularity and precision at the individual level.

2.1 Database Development

Expanding beyond national nutrition surveillance, several institutions have undertaken initiatives to harmonize nutrition data from countries worldwide. The aim is to integrate nutrition surveillance systems into a global context, facilitating the assessment of inequities and disparities on a global scale. In these global databases, raw data was gathered from various countries, and robust data quality evaluation and statistical methods were employed to generate globally comparable data based on the available information. These databases are often stored and shared in the open-access platform. For example, the WHO Global Database on Child Growth and Malnutrition compiles standardizes, and disseminates child anthropometric data of nutrition surveys conducted globally based on the prevalence below and above-defined cut-off points for four anthropometric indexes, namely weight-for-height, height-for-age, weight-for-age and body mass index (BMI)-for-age, in preschool children are presented using z-scores based on the WHO Child Growth Standards [23]. Similarly, the Global Dietary Database (GDD) is an international collaborative effort to produce comprehensive and comparable estimates of dietary intakes of major foods and nutrients globally. The latest cycle GDD model incorporated 1248 dietary surveys representing 188 countries and 99% of the global population in the years 1990, 1995, 2000, 2005, 2010, 2015, and 2018 [24]. The Global Nutrition and Health Atlas (GNHA) database collected diet-related health information from more than 190 countries from more than 15 public databases. GNHA database covers 20 themes and 1000 indicators within the six dimensions of demographics, dietary Intake, nutritional status, health status, health economics, and food sustainability at the global, regional, and national levels, spanning 30 years [25].

In addition to the efforts to develop global nutrition data repositories, electronic health records (EHRs) are starting to be integrated into population health surveillance systems, which were traditionally confined to clinical practice. This shift is driven by the growing volume of data within EHRs and, crucially, their potential to provide real-time access and timely reporting of patient information, encompassing comprehensive health indicators along with other demographic data. The ease of transferring and merging data between systems further contributes to their integration into broader public health initiatives [26, 27]. It is inevitable that nutrition surveillance, similar to the public health surveillance system, will eventually incorporate data related to nutrition and health from EHRs. This data integration with data from other domains broadens the scope and enhances the effectiveness of monitoring the nutrition system. Linking nutrition data with health and disease data through EHRs will facilitate the identification of a greater number of individuals at risk and pinpoint vulnerable populations with greater precision.

2.2 Technology Advancement

Satellite and remote sensing technologies can be utilized to monitor a variety of conditions, such as climate change, conflict, crisis, drought, floods, and even population movements and infrastructure, which can ultimately impact crop production as well as the population's access to food [28, 29]. The incorporation of data obtained through remote sensing will be a great addition to the nutrition surveillance system, especially for monitoring the food security status within specific regions. Mobile phones have also been used to enhance nutrition surveillance by expediting data collection, transfer, analysis, and dissemination. Additionally, they enable the collection of data at a higher frequency, which can turn nutrition surveillance into a real-time monitoring [30]. More recently, wearable devices have emerged as valuable tools for nutrition monitoring with the integration of artificial intelligence techniques. These devices have the capability to measure crucial nutrition-related vital signs, including blood glucose, blood pressure, personal weight, and physical activities [31]. Additionally, innovative devices have been developed to track nutrition and diet habits. Some wearable devices can now monitor total energy intake, dietary patterns, and the time intervals between meals; other wearable sensors are designed to analyze nutrients and metabolites present in saliva, sweat, and tears [32]. This ongoing development of wearable devices, focusing on the measurement of individual chemical and physical nutrition, is transforming nutrition surveillance into a personalized and precise monitoring experience [33, 34].

In addition to digital transformation, omics technology also brings new opportunities in nutrition surveillance. Transcriptomics, proteomics, and metabolomics have served as a crucial technology for examining the impact of diet on an individual's health, offering a nuanced understanding through the identification of food-derived biomarkers and revealing variations in how individuals metabolize the same foods under diverse health and disease conditions [35]. Specifically, metabolomics can contribute to the discovery of novel biomarkers for dietary intake, the exploration of diet-related diseases via cohort studies, and the evaluation of dietary interventions by analyzing metabolic patterns [36]. The targeted application of metabolomics elevates the accuracy of dietary assessments, offering more precise and actionable data in understanding the variations in dietary intake and nutrition status among the population [37, 38]. This precision thereby enhances the effectiveness and precision of nutrition surveillance, empowering policymakers to implement more targeted actions and policies based on insights gained from nutrition metabolomics.

The future of nutrition surveillance is evolving into a sophisticated data-centered system fueled by the integration of diverse data sources and rapid technological advancements. While contributing diverse information, this transition will introduce complexities in terms of harmonization, sharing, and real-time analysis. Addressing these challenges requires innovative approaches to decipher, analyze, and translate vast and varied information into actionable insights. Navigating the intricacies of data integration becomes crucial for achieving more precise and efficient monitoring of global health and nutrition in this dynamic landscape.

3 Growing Opportunities for Nutrition Data Dashboards

This section explores how visual analytics can turn challenges in nutrition surveillance into opportunities. Specifically, we center our discussion on the pivotal role of nutrition and health dashboards, demonstrating their capacity to effectively reveal health and nutrition information along with disparities at various levels. This section also underscores the existing data limitations in dashboards, advocating for the enhancement of dashboard capabilities through the integration of statistical modeling tools.

3.1 Leveraging the Potential: Visual Analytics in Nutrition

Visual analytics, described as the fusion of computational analysis techniques with interactive visualizations for effective understanding, reasoning, and decision-making based on vast and intricate datasets [39], emerges as a promising approach to address data challenges in the nutrition surveillance [40]. Among the various tools, web-based interactive data dashboards stand out as a versatile platform serving both as a means of data dissemination and sharing as well as a tool for visualization and gaining insights [41]. Among the various tools, web-based interactive data dashboards stand out as a versatile platform. The key element in an interactive dashboard is visualization. Visualization enhances the comprehension of complex and large data sets by providing different types of charts and graphs, which require less cognitive load and are more engaging compared to traditional tabular data. Visual aids such as ordered bar charts and highlighted areas in geographical maps can effectively convey key information, which also fosters communication between data providers and diverse stakeholders who receive the information. Furthermore, interactive dashboards have the potential to leverage visualizations to address each of the V's—Velocity, Volume, Value, Variety, Veracity, and Variability—the six key characteristics of the big data [42], This capacity is attributed to its diverse functionality and features. Here is a more detailed breakdown of how interactive dashboards address challenges in the nutrition surveillance system:

- Velocity: The dashboard is constructed using templates, streamlining the process of data upload online and presentation of insightful graphs. The dashboard can even be updated in real-time, which ensures the timely update and dissemination of critical information related to nutrition and health indicators. This form of data sharing and communication is faster than traditional media releases or academic journal publications, facilitating rapid access to vital information for decision-makers.
- Volume: By harnessing the connectivity to databases hosted on servers or in the cloud, the web-based dashboard enables the storage, management, and updating of large databases. By incorporating filters based on time, location, and socio-economic levels, the dashboards are scalable to extract relevant information from databases and varying data volumes presented on the dashboard and ensure a robust system for data management.
- Value: The dashboard goes beyond mere data representation; it provides a descriptive summary of key indicators in specific areas of interest. Statistical models embedded within the platform offer comparable values and unveil intricate relationships between different locations, subpopulations, and various indicators, thereby delivering actionable insights.
- Variety: Recognizing the diverse nature of nutrition surveillance data, the dashboard offers interactive features, multi-panel figures, and an array of visualization types. This diversity of visuals enables effective representation and interpretation of different data types, ensuring a comprehensive understanding of the multifaceted aspects of nutrition-related information.
- Veracity: Ensuring the credibility of data is paramount in nutrition surveillance. The dashboard has the capacity to ensure the veracity of the data by being hosted on accountable institutions, furnishing metadata and information on missing data, collecting original data from reputable and trustworthy institutions, and supplementing detailed documentation on model implementation.

- Variability: Various visualizations on dashboards can help users identify patterns, trends, and outliers within specific indicators. Interactive features such as drop-downs, tooltips, hover-over effects, and clickable elements can help users customize their visuals to focus on relevant aspects of variability in nutrition and health data while also allowing users to explore variability at different levels of granularity.

3.2 Recent Advancement of the Nutrition Dashboard

To evaluate the progress of dashboard development within the nutrition and health surveillance domain, our previous systematic review discovered that numerous organizations, including schools, research institutes, government agencies, international organizations, hospitals, and industries, are actively engaged in the development of interactive dashboards [43]. These dashboards serve the primary objectives of monitoring food fortification progress, tracking malnutrition trends, and disseminating global nutrient and food intake data. In this chapter, we have updated the dashboard review to incorporate more recent dashboards. The summary of these dashboards is provided in Table 1. The majority of the dashboards available for free online search operate at the global level, supplemented by several others focused specifically on the national level.

Examining the visual elements, the current nutrition dashboards we reviewed consistently incorporate maps to illustrate the geographical distribution of specific indicators. In addition, the colors of the map are strategically employed to represent values, while dots highlight focal points, bringing attention to areas of concern. Bar charts are commonly applied to rank nutrition indicators among diverse comparison groups, shedding light on variations across different subpopulations. Line charts are employed to depict trends over time to present the evolving disparities. Interactive filters enable the focused examination of subpopulations, uncovering nuanced patterns. Furthermore, to specifically identify gaps, scorecard-type figures are utilized for comparing current values against predetermined targets. Structurally, these dashboards typically feature a map showcasing all included geographical areas, with filters enabling the isolation of individual indicators. Following this, national or subnational profiles are provided to offer a detailed exploration of key nutrition and health indicators at granular locations.

In summary, the nutrition dashboard, distinguished from traditional tabular data formats, proves to be a powerful tool for swiftly conveying key messages to a broad audience. This not only facilitates the rapid identification of the severity, trends, and anomalies within the current nutrition surveillance but also enables the pinpointing of subpopulations in urgent need of assistance. Consequently, this aids policymakers in making more targeted, precise decisions to address nutrition issues, contributing to more effective and equitable interventions. The organized structure of these dashboards ensures a comprehensive understanding of nutritional surveillance, presented in visually impactful formats that enhance communication and highlight disparities for informed action.

Table 1 Summary review of nutrition dashboards

Dashboard name	Nutritional content	Host organization	Geographical level	Population disaggregation	Year of latest data
Foods system dashboard	Indicators across food system: policies, food supply, food environments, diets and nutrition	GAIN and the Columbia Climate School	Global and national level	Residence, gender	2021
Global fortification data exchange	Fortification, vitamins and minerals intake, and food availability	Joint effort by the food fortification initiative, global alliance for improved nutrition, iodine global network, and micronutrient forum	Global and national	None	2020
Global dietary database	Individual food and nutrient intakes	Tufts University	Global and national	Age, gender, residence. education level	2018
Local burden of disease	Stunting, wasting, overweight underweight, exclusive breastfeeding	Institute for Health Metrics and Evaluation at University of Washington	Global, national, sub-national, 5 km by 5 km level	Age, gender	2019
Global individual food consumption data tool	Food consumption, dietary diversity, nutrition, food safety	Food and Agriculture Organization of the United Nations	Global and national, and sub-national	Age, gender, residence	2021
Hunger map	Prevalence of insufficient food consumption, food security, acute malnutrition (wasting), chronic malnutrition (stunting)	World Food Programme	Global and national, and sub-national	None	2024

(continued)

Table 1 (continued)

Dashboard name	Nutritional content	Host organization	Geographical level	Population disaggregation	Year of latest data
The state of acute malnutrition	Global acute malnutrition, severe acute malnutrition, Stunting	No wasted lives coalition	Global and national	None	2020
Nutrition landscape information system	Malnutrition, vitamin and mineral deficiency, food security, breastfeeding	World Health Organization	Global and national	None	2020
Vitamin A supplementation interactive dashboard	Vitamin supplementation coverage	United Nations Children's Fund	Global, regional, and national	None	2017
Joint child malnutrition estimates 2021 edition—interactive dashboard	Stunting, overweight, wasting,	United Nations Children's Fund	Global and regional level	None	2020
Global target tracking tool	Child stunting, wasting, and underweight, low birth rate, exclusive breastfeeding, anemia	World Health Organization	Global and national	Age, gender	2019
National Institute of Nutrition Atlas in India	Stunting, wasting, underweight, overweight	India Council of Medical Research—National Institute of Nutrition, Hyderabad	National, state and district level	Age, gender	2015
Global nutrition and health atlas	Demographics, dietary intake, nutritional status, health status, health economics, and food sustainability	Tufts University	Global and national	Age, gender, residence	2021

(continued)

Table 1 (continued)

Dashboard name	Nutritional content	Host organization	Geographical level	Population disaggregation	Year of latest data
Standardized expanded nutrition survey data dashboard	Child malnutrition, anemia, food security, infant feeding, water sanitation	UN Refugee Agency	Global and national	Age, gender	2023
Global Food and Nutrition Security Dashboard	Food security indicators	Global Alliance for Food Security	Global and national, and sub-national	Age, gender	2023
NHANES interactive data visualizations	Dietary intake, nutrition status	Centers for Disease Control and Prevention in United States	National	Age, gender, race	2018
Global diet quality project	Food group diversity score and food group consumption	Collaboration between Gallup, Harvard Department of Global Health and Population, and the Global Alliance for Improved Nutrition (GAIN)	Global and national	Gender, residence	2023
Multisectoral nutrition dashboard in Bangladesh	Nutrition situation and interventions	Bangladesh National Nutrition Council	National, division and district level	Age, gender, residence	2018
Food security and nutrition dashboard in Kenya	Food security, nutrition outcomes, nutrition drivers	National Information Platform for Food Security Nutrition	National and county level	Age, gender, residence	2022

3.3 Navigating Present Challenges in Nutrition Dashboards

There are also some common challenges with dashboards. One prevalent challenge associated with dashboards is the limitation in data granularity. The aggregation of data is a common practice in dashboards that encompass information from various countries, often resulting in a lack of granular data availability or the need for aggregation to provide a concise summary. However, data aggregation on the dashboard may obscure disparities across demographical and geographical subpopulations. When data is aggregated based on demographic factors such as age and sex, it can mask the distinct patterns of nutrition and health indicators within subgroups [44]. Moreover, due to the lack of a standardized scheme, data aggregation is often carried out with different rules and thresholds, making it challenging to compare results and synchronize data across different studies. Spatial aggregation, similar to other aggregation data, typically operates under the assumption of uniformity across sub-level geographical dimensions. This assumption can lead to overlooking critical nuances, such as localized hotspots [45], where specific nutritional challenges may be concentrated. Additionally, the clustering of nutrition indicators may not be solely due to geographical proximity but could also be influenced by factors such as access to sentinel sites, transportation networks, or the distribution of natural resources, which further biases what is included in the data in the first place [46–48].

Simply using line charts to track the average trends in nutrition indicators over time might not be sufficient to fully understand the complex and dynamic nature of nutrition status longitudinally. For example, the temporal changes in nutrition and health outcomes could be very complex and interconnected with many factors, such as the seasonal fluctuations of harvesting [49], social calendars [50], long-term food security [51], climate change [52], economic recession [53], and emerging infections [54]. However, limited data collection on a variety of indicators and statistical model explorations have hindered the examination of complex temporal relationships. It is worth noting that dietary intake data is often reported on the dashboard as the mean or median value for a given country instead of the full distribution [55]. However, relying solely on the central tendency of dietary intake can be problematic, especially when the distribution is unknown. This approach can lead to biased results and can overlook the spatial variations in the mean and the full picture of the intake distribution. Understanding these complexities is of paramount importance as it underscores the underlying patterns and disparities within the data.

In our systematic review of dashboards and deployment of the Global Nutrition and Health Atlas, we emphasized the need for comprehensive evaluation and scoring platforms integrating data portals and visualization dashboards [25, 43]. We introduced the concept for evaluating and developing data visuals based on a set of well-grounded principles and metrics. These four principles, Evidence, Efficiency, Emphasis, and Ethics (4Es), form rules to construct visuals, organize information, and communicate science to broad audiences. In brief:

- Evidence: This principle relies on the content, results, or concepts that correspond to the information the graph conveys, and those logic, content, results, or concepts are transferrable to a verbal description. This principle stipulates that graphs should be created with a well-thought-out and well-grounded description in mind.
- Efficiency: This principle states that a graph should help explain the results or concepts by taking advantage of visual perception. Thus, a graph should utilize good visual properties with respect to culture-driven and physiology-driven visual perceptions.
- Emphasis: This principle states that a graph should be user-friendly, force the audience to note the unexpected, motivate questions, and clarify statements. Thus, a graph should be created with cognizance of the intended audience.

- Ethics: This principle involves achieving the truth and promoting essential values such as accuracy, precision, reliability, integrity, and inclusiveness at neither excess nor deficiency but at the mean of these two. Thus, a graph should be created with awareness of its perception by the intended audience.

These principles are adaptable beyond individual visualizations; we adapted and applied these principles to evaluate interactive platforms integrating data portals and visualization dashboards utilized to describe, monitor, and track records temporally and spatially. We generated a set of 48 evaluation metrics for evaluating the dashboard, including if the stated goals of a dashboard matched their actual content, the data and visualization quality, the capability, comprehension, and accessibility of the dashboard, etc., of the dashboard based on 4 major principles. For example, we formulated seven metric questions to evaluate the data granularity of a dashboard, which is a part of the data quality evaluation. Specifically, the metric questions asked first if the dashboard clarified their level of aggregations on spatial, temporal, demographical, and/or nutrition and health characteristics and further asked if the dashboard was able to display those data by disaggregating into different temporal and geographical and/or nutrition indicators levels. We then conducted a systematic review of 13 dashboards utilizing a set of 48 evaluation metrics for data integrity, completeness, granularity, visualization quality, and interactivity. We summarized the dashboards' characteristics, strengths, and limitations from a common user perspective, and applied mixed effects models to summarize evaluation results adjusted for inter-rater variability and provided guidelines for nutrition dashboard development. After evaluating dashboards using the proposed metrics and evaluation principles, we found that the biggest gap among current nutrition and health dashboards lies in data standardization and harmonization, particularly with regard to geo-temporal granularity, as well as reporting on missing data.

Lastly, analytical functions are underutilized in nutrition dashboards. While dashboard creators frequently showcase modeled data, the absence of analytical functions like statistical testing, association parameters, or regression models limits our ability to comprehensively examine and understand nutrition outcomes, particularly with disparities among different population groups. The integration of powerful visual analytics into nutrition surveillance shouldn't be confined to basic graphical representations; rather, the inclusion of advanced statistical models and results within the dashboard can significantly augment its potency and value to unveil hidden patterns, illuminate disparities, and pinpoint areas of inequity. Incorporating analytical functions into the nutrition dashboard incorporating analytical functions, nutrition dashboards can transcend mere data visualization and is also a pivotal step towards unlocking the full potential of these tools in advancing research, policy formulation, and targeted interventions for improved public health outcomes.

4 Conclusions

Nutrition surveillance systems combined with data sharing and democratization implemented in integrative data dashboards are critical for boosting information and knowledge sharing among researchers, practitioners, and decision-makers in the field of food and nutrition and help to accelerate data literacy and science communication. With growing novel analytical tools and computational capacities, leaders and citizens of our interconnected world should be well-equipped to tackle complex problems and answer big questions, such as: How can we identify and tackle health disparities using advanced technologies? Can we prevent and mitigate health disparities induced by global challenges? How can we better communicate scientific facts, risks, and uncertainties to the public about complex problems related to but limited to food and nutrition? These questions are not new, yet they remain pressing and critical. With growing interest in artificial intelligence (AI), precision public health, and nutrition, we now wonder who is responsible for AI-generated policies and how to create transparent and inclusive rules for data collection, sharing, and communication [56–58] and we should be ready to answer to these questions to advance public health to our high aspirations.

References

1. Report of the World Food Conference, Rome, 5–16 November 1974 (1975) United Nations. https://digitallibrary.un.org/record/701143?ln=en
2. World Health Organization (1976) Methodology of nutritional surveillance: report of a Joint FAO/UNICEF/WHO Expert Committee [meeting held in Geneva from 1 to 10 October 1975]. World Health Organization
3. Mason JB, Mitchell JT (1983) Nutritional surveillance. Bull World Health Organ 61(5):745
4. Ismail SJ (1991) Nutritional surveillance: experiences from developing countries. Proc Nutr Soc 50(3):673–679
5. Al Jawaldeh A, Osman D, Tawfik A, World Health Organization (2013) Food and nutrition surveillance systems: technical guide for the development of a food and nutrition surveillance system for countries in the Eastern Mediterranean Region
6. Organization WH (2017) Global nutrition monitoring framework: operational guidance for tracking progress in meeting targets for 2025. https://apps.who.int/iris/bitstream/handle/10665/259904/9789246513604-ara.pdf
7. Haddad L, Achadi E, Bendech MA, Ahuja A, Bhatia K, Bhutta Z, Blössner M, Borghi E, Colecraft E, De Onis M (2015) The Global Nutrition Report 2014: actions and accountability to accelerate the world's progress on nutrition. J Nutr 145(4):663–671
8. Organization WH (2019) Nutrition landscape information system (NLiS): results of a user survey. World Health Organization. https://apps.who.int/iris/bitstream/handle/10665/328980/WHO-NMH-NHD-19.10-eng.pdf
9. Al Jawaldeh A, El Hajj Hassan O, Qureshi AB, Zerbo FC, Alahnoumy S, Bozo M, Al-Halaika M, Al-Dakheel MH, Alhamdan L, Mujib SA (2023) Qualitative review of national nutrition surveillance systems in the Eastern Mediterranean Region. Nutrients 15(17):3689
10. Peters R, Li B, Swinburn B, Allender S, He Z, Lim SY, Chea M, Ding G, Zhou W, Keonakhone P (2023) National nutrition surveillance programmes in 18 countries in South-East Asia and Western Pacific Regions: a systematic scoping review. World Health Organization. Bull World Health Organ 101(11):690–706

11. The ELANS Study Group, Fisberg M, Kovalskys I, Gómez G, Rigotti A, Cortés LY, Herrera-Cuenca M, Yépez MC, Pareja RG, Guajardo V, Zimberg IZ, Chiavegatto Filho ADP, Pratt M, Koletzko B, Tucker KL (2015) Latin American Study of Nutrition and Health (ELANS): rationale and study design. BMC Public Health 16(1):93. https://doi.org/10.1186/s12889-016-2765-y
12. Breda J, Castro LSN, Whiting S, Williams J, Jewell J, Engesveen K, Wickramasinghe K (2020) Towards better nutrition in Europe: evaluating progress and defining future directions. Food Policy 96:101887
13. Rippin HL, Hutchinson J, Evans CE, Jewell J, Breda JJ, Cade JE (2018) National nutrition surveys in Europe: a review on the current status in the 53 countries of the WHO European region. Food Nutr Res 62. https://www.ncbi.nlm.nih.gov/pmc/articles/PMC5917420/
14. Tuffrey V, Hall A (2016) Methods of nutrition surveillance in low-income countries. Emerg Themes Epidemiol 13(1):1–21
15. Backer D, Billing T (2021) Validating famine early warning systems network projections of food security in Africa, 2009–2020. Glob Food Sec 29:100510
16. Daher J, Cassard F, Gari S, Sauveplane-Stirling V, Nyawo M, Codjia P (2018) Implementation of nutrition surveys using SMART methodology in sub-Saharan Africa. Field Exchange 58:68
17. Briefel RR (1994) Assessment of the US diet in national nutrition surveys: National collaborative efforts and NHANES. Am J Clin Nutr 59(1):164S-167S
18. Byers T, Sedjo RL (2012) Nutrition monitoring and surveillance. Nutr Epidemiol 344
19. Erdman JW, Macdonald IA, Zeisel SH (eds) (2012) Present knowledge in nutrition, 1st edn. Wiley. https://doi.org/10.1002/9781119946045
20. Rivera JA, Sepúlveda Amor J (2003) Conclusions from the Mexican National Nutrition Survey 1999: translating results into nutrition policy. Salud Pública de México 45(suppl 4):565–575
21. Guibrunet L, Ortega-Avila AG, Arnés E, Mora Ardila F (2023) Socioeconomic, demographic and geographic determinants of food consumption in Mexico. PLoS ONE 18(10):e0288235
22. Canada S (2017) 2015 Canadian Community Health Survey (CCHS)–nutrition: user guide. Statistics Canada Ottawa, ON, Canada
23. De Onis M, Blossner M, W. H. Organization (1997) WHO global database on child growth and malnutrition. World Health Organization. https://apps.who.int/iris/bitstream/handle/10665/63750/WHO_NUT_97.4.pdf
24. Miller V, Cudhea F, Singh G, Micha R, Shi P, Zhang J, Onopa J, Karageorgou D, Webb P, Mozaffarian D (2019) Estimated global, regional, and national cardiovascular disease burdens related to fruit and vegetable consumption: an analysis from the global dietary database (FS01-01-19). Curr Dev Nutr 3:3130601
25. Zhou B, Liang S, Monahan KM, El-Abbadi N, Cruz MS, Chen Y, DeVane A, Reedy J, Zhang J, Semenova I (2022) An open-access data platform: Global Nutrition and Health Atlas (GNHA). Curr Dev Nutr 6(4):nzac031
26. Paul MM, Greene CM, Newton-Dame R, Thorpe LE, Perlman SE, McVeigh KH, Gourevitch MN (2015) The state of population health surveillance using electronic health records: a narrative review. Popul Health Manag 18(3):209–216. https://doi.org/10.1089/pop.2014.0093
27. Birkhead GS, Klompas M, Shah NR (2015) Uses of electronic health records for public health surveillance to advance public health. Annu Rev Public Health 36(1):345–359. https://doi.org/10.1146/annurev-publhealth-031914-122747
28. Brown ME (2016) Remote sensing technology and land use analysis in food security assessment. J Land Use Sci 11(6):623–641. https://doi.org/10.1080/1747423X.2016.1195455
29. Karthikeyan L, Chawla I, Mishra AK (2020) A review of remote sensing applications in agriculture for food security: crop growth and yield, irrigation, and crop losses. J Hydrol 586:124905
30. Barnett I, Gallegos JV (2013) Using mobile phones for nutrition surveillance: a review of evidence. https://opendocs.ids.ac.uk/opendocs/handle/20.500.12413/2602
31. Dias D, Cunha PS (2018) Wearable health devices—vital sign monitoring, systems and technologies. Sensors 18(8):2414

32. Sempionatto JR, Montiel VR-V, Vargas E, Teymourian H, Wang J (2021) Wearable and mobile sensors for personalized nutrition. ACS Sens 6(5):1745–1760. https://doi.org/10.1021/acssensors.1c00553
33. Shi Z, Li X, Shuai Y, Lu Y, Liu Q (2022) The development of wearable technologies and their potential for measuring nutrient intake: towards precision nutrition. Nutr Bull 47(4):388–406. https://doi.org/10.1111/nbu.12581
34. Mortazavi BJ, Gutierrez-Osuna R (2023) A review of digital innovations for diet monitoring and precision nutrition. J Diabetes Sci Technol 17(1):217–223. https://doi.org/10.1177/19322968211041356
35. Corella D, Ordovás JM (2015) Biomarkers: background, classification and guidelines for applications in nutritional epidemiology. Nutr Hosp 31(3):177–188
36. Naska A, Lagiou A, Lagiou P (2017) Dietary assessment methods in epidemiological research: Current state of the art and future prospects. F1000 Research 6. https://www.ncbi.nlm.nih.gov/pmc/articles/PMC5482335/
37. Tebani A, Bekri S (2019) Paving the way to precision nutrition through metabolomics. Front Nutr 6:41
38. Brennan L, Hu FB, Sun Q (2021) Metabolomics meets nutritional epidemiology: harnessing the potential in metabolomics data. Metabolites 11(10):709
39. Cui W (2019) Visual analytics: a comprehensive overview. IEEE Access 7:81555–81573
40. Ola O, Sedig K (2014) The challenge of big data in public health: an opportunity for visual analytics. Online J Publ Health Inform 5(3):223. https://doi.org/10.5210/ojphi.v5i3.4933
41. Manorat R, Becker L, Flory A (2019) Global data visualization tools to empower decision-making in nutrition. Sight and Life 33(1):108–114
42. Al-Sai ZA, Abdullah R, Husin MH (2019) Big data impacts and challenges: a review. IEEE Jordan Int Joint Conf Electr Eng Inform Technol (JEEIT) 2019:150–155. https://doi.org/10.1109/JEEIT.2019.8717484
43. Zhou B, Liang S, Monahan KM, Singh GM, Simpson RB, Reedy J, Zhang J, DeVane A, Cruz MS, Marshak A (2022) Food and nutrition systems dashboards: a systematic review. Adv Nutr 13(3):748–757
44. Diaz T, Strong KL, Cao B, Guthold R, Moran AC, Moller A-B, Requejo J, Sadana R, Thiyagarajan JA, Adebayo E (2021) A call for standardised age-disaggregated health data. Lancet Healthy Longevity 2(7):e436–e443
45. Smith DM, Cummins S, Taylor M, Dawson J, Marshall D, Sparks L, Anderson AS (2010) Neighbourhood food environment and area deprivation: Spatial accessibility to grocery stores selling fresh fruit and vegetables in urban and rural settings. Int J Epidemiol 39(1):277–284
46. Jia P, Wang Y, Yang M, Wang L, Yang X, Shi X, Yang L, Wen J, Liu Y, Yang M (2022) Inequalities of spatial primary healthcare accessibility in China. Soc Sci Med 314:115458
47. Striessnig E, Bora JK (2020) Under-five child growth and nutrition status: spatial clustering of Indian districts. Spat Demography 8:63–84
48. Cordeiro LS, Otis NP, Sibeko L, Nelson-Peterman J (2021) Rural-urban disparities in the nutritional status of younger adolescents in Tanzania. PLoS ONE 16(12):e0261480
49. Stelmach-Mardas M, Kleiser C, Uzhova I, Peñalvo J, La Torre G, Palys W, Lojko D, Nimptsch K, Suwalska A, Linseisen J (2016) Seasonality of food groups and total energy intake: a systematic review and meta-analysis. Eur J Clin Nutr 70(6):700–708
50. Hasan MM, Kader A, Asif CAA, Talukder A (2023) Seasonal variation in the association between household food insecurity and child undernutrition in Bangladesh: mediating role of child dietary diversity. Matern Child Nutr 19(2):e13465
51. McQuade ETR, Clark S, Bayo E, Scharf RJ, DeBoer MD, Patil CL, Gratz JC, Houpt ER, Svensen E, Mduma ER (2019) Seasonal food insecurity in Haydom, Tanzania, is associated with low birthweight and acute malnutrition: Results from the MAL-ED study. Am J Trop Med Hyg 100(3):681
52. Brown ME, Backer D, Billing T, White P, Grace K, Doocy S, Huth P (2020) Empirical studies of factors associated with child malnutrition: highlighting the evidence about climate and conflict shocks. Food Secur 12:1241–1252

53. De Vogli R, Owusu JT (2015) The causes and health effects of the great recession: from neoliberalism to 'healthy de-growth.' Crit Public Health 25(1):15–31
54. Katona P, Katona-Apte J (2008) The interaction between nutrition and infection. Clin Infect Dis 46(10):1582–1588
55. Micha R, Coates J, Leclercq C, Charrondiere UR, Mozaffarian D (2018) Global dietary surveillance: data gaps and challenges. Food Nutr Bull 39(2):175–205
56. Naumova EN (2023) Who is responsible for AI-generated public health policies? J Public Health Policy 44(4):517–522. https://doi.org/10.1057/s41271-023-00438-1
57. Naumova EN (2022) Precision public health: is it all about the data? J Public Health Policy 43(4):481–486. https://doi.org/10.1057/s41271-022-00367-5
58. Naumova EN (2024) Telling truth with data visuals: a guide for public health professionals. J Public Health Policy 45(2):1–7. https://doi.org/10.1057/s41271-024-00479-0

Addressing Health Equity Challenges in Women's Care: Insights from Kenya

11

E. Akosa Wambalaba

Abstract

This chapter identifies and discusses socio-economic and cultural barriers to health care among low-income and rural women in Kenya. Its focus is on understanding and highlighting the concept of personhood, and how it intersects with the contextual economic realities, cultural beliefs and practices, and social expectations in shaping women's behavior towards health care and promoting health care inequity. The chapter shows that healthcare facilities tend to use interventions from high-income areas, "one size fits all" and overlook the important role culture and communities exert on women, especially rural women, and women of low economic status in making health decisions. It also highlights personal attributes such as bias, cultural and professional incompetence, and financial status as contributors to inadequate care and the widening health inequity gap. Existing barriers to care are identified and resolutions are proposed to narrow the gap between intervention and implementation. Incorporating understanding and proficiency in the cultural and social expectations surrounding women's health-seeking behavior, alongside local community, institutional, and healthcare provider knowledge, will result in care that is rooted in, collaborates with, and serves the needs of the communities and women it serves. The chapter summarizes how socioeconomic and cultural barriers impact prevention, diagnosis, treatment, and cure. Specific intervention strategies that take into account local cultural realities affirm the influence of community. For a deeper understanding, a comprehensive conceptual framework is designed on the relationship between women's health decision-making and health inequity.

E. A. Wambalaba (✉)
Department of General Education and Department of Languages and Literature, United States International University-Africa, Nairobi, Kenya
e-mail: awambalaba@usiu.ac.ke

B. W. K. Son (ed.), *The Landscape of Global Health Inequity*, Integrated Science 22,
https://doi.org/10.1007/978-3-031-60502-4_11

Keywords

Care • Socioeconomic barriers • Cultural barriers • Intervention • Rural women • Social expectations • Health decision-making

1 Introduction

Vast evidence exists on various drivers of healthcare disparities globally [1]. In developing countries such as Kenya, where social structures are predominantly influenced by communal practices and beliefs, the cumulative impact of socio-economic and cultural barriers imposes a substantial burden of harm on the health of women and girls. Young women and adolescent girls, in particular, are vulnerable to poverty, gender-based violence, harmful cultural attitudes and beliefs, and limited control over benefits from land and other resources which consequently impair their healthcare outcomes, aggravating health inequities [2, 3].

Socioeconomic and cultural factors generally impact access and utilization of health services and the type and quality of care received at health facilities [4]. This is especially true for women in rural areas and low-income women in urban areas of resource-constrained developing countries like Kenya. This leads to inequity, which simply means giving people, organizations, and communities what they need to be successful [5]. Reducing care inequity influenced by detrimental socio-economic and cultural factors will enhance women's well-being through increased utilization and enhanced quality of care. Therefore, comprehending the intersection of women's decision-making processes, social, economic, and cultural circumstances with access to care, utilization patterns, and care quality is essential for crafting effective interventions for policymakers and healthcare planners.

Though it is widely acknowledged that cultural and socio-economic factors impact healthcare equity, these factors vary and have varied implications on healthcare in different settings. Socioeconomic factors refer to the interrelatedness of variables like education, income, security, employment status, social status, social expectations and responsibility, residence location, and religion. Cultural factors point to the practices, beliefs, attitudes, expectations, and values associated with one's social identity such as ethnicity, race, and so on, or the dominant accepted cultural practices and beliefs in a localized area [4]. Both socio-economic and cultural factors can potentially impede or facilitate the access and utilization of healthcare.

Of interest in this chapter, is how an aggregate of all the above-mentioned socio-economic and cultural factors intersects with gender to sustain or facilitate healthcare inequity for low-income women in urban areas and those in rural areas in Kenya. In contrast to some other developing nations where barriers to healthcare access primarily stem from income or race, Kenya faces unique challenges deeply ingrained within its cultural and social fabric. Dominant religious and cultural practices, education levels, income disparities, and social expectations significantly influence health outcomes related to family planning, pregnancy, female morbidity, and mortality in the country [6]. This is especially the case in many parts of rural Kenya, where local ethnic cultures and religions are dominant and hold strong influence over the beliefs, attitudes, and behavior of members. Rural societies are collectively oriented and social expectations and responsibilities hold great sway and non-conformity is accompanied by socially isolating sanctions [7].

Given the important influence of socio-economic and cultural influences on women's health care in Kenya, it is crucial to explore and identify individual socio-economic and cultural barriers to find solutions that fit within the structure of social expectations and responsibilities of a communal environment [8]. Understanding and unraveling sociocultural influences necessitates an exploration of the concept of personhood as culturally constructed within many Kenyan communities. Examining research that has identified successful interventions in low-resource countries, particularly those with large populations of women in rural areas constrained by tradition, culture, social expectations, and low incomes, both in urban and rural settings, offers potential resolutions to enhance care and diminish the health equity gap for women in Kenya. This approach bridges the divide between intervention and implementation, fostering more effective healthcare practices.

The objective of this study is to formulate a framework that comprehensively elucidates women's healthcare in Kenya as a multifaceted issue entrenched within community dynamics, surpassing the influences of economics or education alone. Additionally, it draws upon potential remedies from existing studies to enhance women's care and narrow the gender equity gap. The subsequent sections of this chapter are structured as follows. The following section furnishes an overview of the concept of personhood in predominantly communal societies like Kenya and delineates how the interpretation of personhood shapes social interactions [9]. Subsequently, there is an examination of various socio-economic and cultural barriers to care identified in the literature, along with their impacts and proposed resolutions. Finally, a conceptual framework illustrating the relationship between women's health decisions and health equity is presented.

2 The Concept of Personhood

To gain deeper insight into the nature of African women's engagement with health, it is crucial to grasp the concept of personhood as a fundamental driver. Personhood represents the entity that interacts with the healthcare system, shaping decisions, problem-solving capabilities, values, and beliefs based on socialized learning. It embodies a moral agent whose behaviors are evaluated within the moral standards or essential criteria of personhood established by their community. This framework underscores the significance of understanding personhood in elucidating African women's health dynamics [10]. This suggests that an individual is perceived as a moral agent with agency, accountable for their actions within the context of cultural expectations and roles. Within this framework, individuals are responsible for navigating their behaviors within the established norms and values of their culture, reflecting their roles and responsibilities within the community. The concept of personhood exhibits cross-cultural variation, with each cultural context shaping its distinctive idea rooted in its values, practices, and systems of thought, considered "normal" within that framework [10]. In many African cultures, personhood is centered around relationships, emphasizing interconnectedness in three primary dimensions: Firstly, the spiritual relationship involves a person's connection with God, myths, spirits, or the power of ancestors. Secondly, the social relationship encompasses connections with family, friends, clan, ethnic group, and humanity at large. Lastly, one's relationship with the self is emphasized, reflecting an internal understanding and connection with one's identity and purpose [9].

In communal communities, socioeconomic and cultural factors can give rise to both tangible and intangible barriers, deeply ingrained in systemic, gendered cultural perceptions of health conditions, attitudes, and practices. These barriers can discriminate against and disadvantage women in healthcare-seeking behavior in two primary ways: Firstly, they influence the choice of healthcare facilities frequented, and secondly, they impact the type of treatment sought or received [6]. The relationship between women and health is thus approached similarly, with individuals attributing meaning to situations and shaping decision-making through a triad of relational connectedness, influenced by cultural notions of personhood and self-representation. Within this relational triad, agency is anticipated to conform to social expectations and garner approval. Therefore, it is essential to consider contextual notions of personhood and their degree of influence on women's health-seeking behavior to gain a deeper understanding of underlying causes and to design effective interventions.

3 Socioeconomic Barriers to Care and Resolutions

3.1 Affordability

Social determinants of health care include socio-economic status, education, living environment, employment, social support networks, and access to care [4]. Addressing social determinants of health is important in improving health and reducing disparities in health care. Kenya has a population of 47 million and is categorized as a lower-middle-income country [11]. Despite being a lower middle-income country, poverty levels are still chronically high for the majority of the population, impacting the cost of care that remains high and unaffordable for the majority of women [1, 12]. Figure 1 outlines a conceptual framework on affordability as a barrier to women's care in Kenya and resolutions.

Gender-based income disparities persist at alarming rates in Kenya, exerting profound effects on health outcomes and exacerbating healthcare inequities. According to the 2022 Demographic and Health Survey conducted by the Kenya National Bureau of Statistics, women across all age groups earn an average of

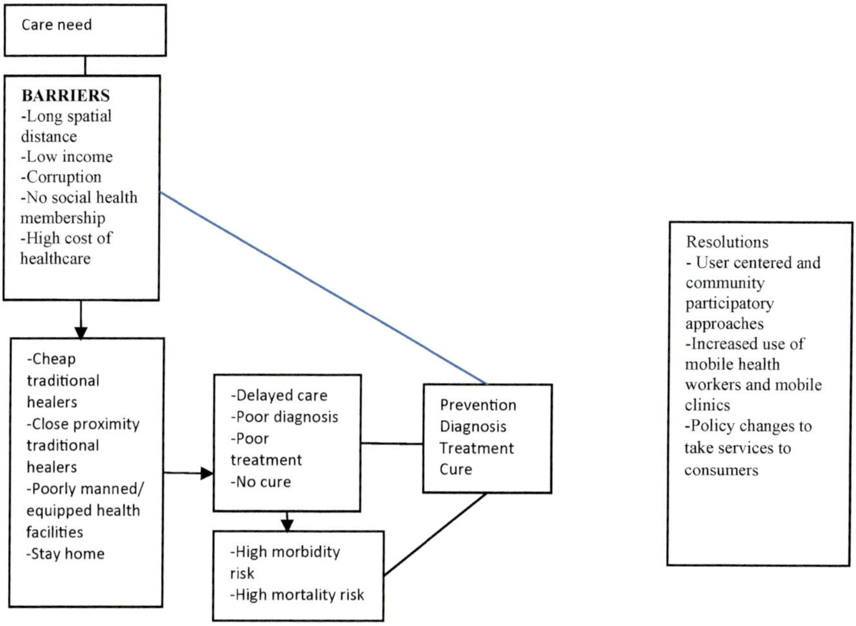

Fig. 1 Conceptual framework on affordability as a barrier to women's care in Kenya

Kshs. 12,166 per month, in contrast to men's average of Kshs. 18,595 [7]. Furthermore, women earn 30% less than men in the same job categories [13]. The prevalence of poverty and lower salaries among women perpetuates disproportionate income levels compared to males, significantly impacting their access to and the quality of healthcare they receive [14]. The limited earning power among women directly affects their capacity to exercise agency in matters related to health decisions and seek quality healthcare without being restricted to poor services provided in publicly operated facilities.

Kenya has public and private health insurance schemes. However high poverty levels continue shutting women out of insurance schemes. In 2022, only 1 in 25 Kenyans had health coverage, while 26% of women had either public, private, or community-based health insurance, leaving 74% of the population uninsured [7]. In urban areas, 38.9% of women possess some form of health insurance, in contrast to 19.5% of rural women. However, the overall percentage of women in Kenya with any type of health insurance remains low, standing at 5% for both females and males in the lowest income bracket, and at 56% for females and 60% for males in the upper bracket. The uptake of insurance correlates with household wealth [7]. Notably, 84% of Kenyans covered by insurance rely on government-managed social health insurance, specifically the National Hospital Insurance Fund (NHIF), to access healthcare services.

Social health insurance aims at improving health care and health outcomes by decreasing health care inequities. It also seeks to mitigate the effects of poverty on personal health service access by reducing or eliminating out-of-pocket payments for health services. Despite Kenya having a social health insurance scheme, it charges a monthly membership fee and many Kenyan women cannot afford the family membership fee of Kshs 500 or approximately $3. Many women are not members and cannot benefit from the social health care system [7]. The exorbitant cost of healthcare in Kenya often excludes women from accessing better-equipped public hospitals staffed by professional healthcare providers. Consequently, women are compelled to seek care in under-equipped and understaffed public health centers or resort to the services of untrained traditional healers, who offer cheaper alternatives. A study revealed that 90% of maternal deaths result from inadequate quality of care [1]. In a study focusing on Somali ethnic women in Kenya seeking healthcare for complications related to female genital mutilation (FGM), the elevated cost of healthcare emerged as a crucial determinant influencing the quality of healthcare services sought by women [15]. Women often choose to seek healthcare services from traditional healers due to their affordability and accessibility, resulting in subpar diagnosis and treatment. Public healthcare facilities were typically considered a last resort if the condition persisted untreated or worsened [14].

Affordability, particularly viewed in terms of spatial distance, significantly impacts the demand for healthcare [16]. The considerable physical distance between a woman's residence and the nearest health facility can render motorized transport expensive, unavailable, unreliable, or unaffordable for many rural women. In rural areas, the challenges of accessing healthcare facilities due to their remote location or distance often lead pregnant women to initially seek alternative services from nearby traditional healers or opt for home deliveries [17]. These proximity-based decisions delay women from accessing quality care in better-equipped public health facilities and increase the risk of infant or maternal mortality [7]. This may partly account for Kenya's persistently high rates of infant and maternal mortality. In 2021, the infant mortality rate stood at 28 deaths per 1000 live births, the child mortality rate at 32.7 deaths per 1000 live births, and the maternal mortality ratio was 530 per 100,000 live births [18].

Confirming the trend of delaying the seeking of quality health services, the Kenya Demographic and Health Survey of 2022 revealed that a lower percentage of women, 77% residing in rural areas, delivered in health facilities compared to urban women, at 91.7% [7]. Furthermore, in the same survey, 97% of urban respondents had given birth through a skilled provider, compared to 84% of women in rural areas [7]. Urban women benefit from direct access to a diverse urban transport infrastructure that is quick, affordable, and readily available for accessing health facilities and receiving treatment, unlike the underdeveloped and unreliable transport infrastructure in rural areas. Additionally, the affordability factor is limited in comparatively lower-salary rural areas [19].

Corruption significantly affects the cost of healthcare in Kenya. NHIF membership entitles individuals to free services in public health facilities and requires out-of-pocket payments in private ones, which cover cost differentials if substantially higher than in public facilities. However, despite having health insurance cards, some healthcare providers still demand payment for specialist services or refuse to accept NHIF cards for services [16]. Advocating for the interests of vulnerable groups within communities to engage with health service providers and public institutions may help reduce corruption levels and improve health outcomes [5].

3.2 Social Barriers to Care: Social Capital and Resolutions

Communal-based groupings represent a significant dimension of social connectedness that many women in Kenya rely on for financial support in treating adverse health conditions. Social capital groupings constitute resource-rich social networks that individuals leverage to benefit themselves and their networks, relying heavily on trust and reciprocity [20]. In Nairobi city, social capital holds high value and is deeply ingrained in the relationships of urban low-income women, serving as a safety net reservoir during sudden financial disruptions [6]. One prominent source of social capital accessible to women across all socio-economic brackets is membership in socio-economic groupings known as "chamas" or merry-go-rounds. These groupings serve as both financial and emotional support networks, aligning with cultural representations of connectedness and providing women with a cushion during sudden financial shocks. Additionally, they function as intermittent revenue streams.

Social capital influences healthcare at two levels: preventive care and treatment. The caliber of an individual's social capital shapes the extent of information dissemination and access among its members, thereby influencing the quality of healthcare-seeking decisions and behaviors. A study conducted by Musalia [20] in Kenya unveiled a correlation between trust, a component of social capital, and enhanced health outcomes through storytelling, information exchange, and gatherings within bridging social groups. Bridging social capital proves more advantageous as it fosters an environment of diverse perspectives and information sharing, fostering the accumulation of knowledge, experiential understanding, and support, and facilitating fluid behavioral changes.

Women in rural low-income brackets often encounter operational barriers when attempting to leverage social capital for information and ideas due to the limited scope of information available. This limitation stems from the homogeneous nature of their social networks, which produces non-varied, restricted information, thereby hindering the emergence of autonomy—a precursor to agency. In contrast, women in urban areas experience expanded health-seeking behaviors and access due to their proximity to rich social capital characterized by homogeneity, and diversity in member characteristics, labeled as bridging rather than bonding groupings associated with rural women. This positively impacts their awareness, knowledge, and ability to exercise autonomy, build confidence, and make informed decisions beneficial to their care [20].

3.3 Social Barriers to Care: Language and Literacy Levels and Resolutions

Health literacy encompasses the ability to locate, comprehend, and utilize information and services to inform health-related decisions and actions. This capacity is closely tied to the impact of economic status on access to and the quality of education. In Kenya, enhancing healthcare for women requires increased efforts to bridge the gap between language proficiency and health literacy levels. Poor communication resulting from language differences or significant disparities in literacy levels poses barriers to effective healthcare delivery between healthcare providers and women seeking medical assistance. Low literacy levels among women seeking care impede effective communication, degrading the quality of interaction [15]. Specifically, low literacy levels among Somali women have been identified as a hindrance to effective communication between patients and healthcare providers [15]. The similarity in language, values, traditions, and customs facilitates more open and honest communication between healthcare providers and patients, fostering trust and thereby enhancing diagnostic accuracy and health outcomes [21].

The inability to speak English or the national language Kiswahili in Kenya often correlates with one's economic status, which directly impacts the quality of education received. Dissimilarities in language can pose a significant barrier to accessing quality care for women who associate language similarity with trust. Within the Somali community, language serves as a conduit for culture and belonging, making discussions about sexual health-related matters with a male health provider who speaks a different ethnic language frowned upon and labeled taboo [15]. Differences in language between healthcare providers and patients serve as a major impediment to effective communication, hindering the exchange of crucial information [15]. Consequently, women may find themselves restricted in their ability to receive quality and lower-cost care due to cultural biases, misconceptions, and stereotypes surrounding language barriers.

3.4 Social Barriers to Care: Mobile Phone Technology and Resolutions

In Kenya, mobile phones have become affordable and offer easy and reliable internet connectivity through data bundles, democratizing the space for information gathering. Leveraging mobile phones, which are ubiquitous, portable, reliable, and accessible, presents an opportunity for more equitable health access and care. They can be used to enhance health literacy in the user's language and disseminate information about services, preventative care, and treatment. Additionally, mobile phones can store vital health information, thus overcoming accessibility barriers by facilitating communication between patients and healthcare providers for health-related information or services. Access to a mobile phone can thus mitigate healthcare disparities caused by physical distance or other barriers.

Kenya boasts a high per capita phone ownership, with mobile services being largely affordable. By the end of March 2023, there were 66.1 million mobile phone subscribers in Kenya, reflecting a mobile penetration rate of 130.5% or 122 cell phone subscriptions per 100 people [23]. The Kenya 2019 Census report revealed that women own 50.6% of mobile phones compared to men's 49.8% [11]. The International Telecommunication Union calculated the average monthly phone budget in Kenya for 2020 at $10.30, which is considered affordable, particularly with an income of less than $3.20 per day according to USAID data [1]. Developing, scaling, and enhancing accessibility to mHealth can serve as a potent tool for narrowing the equity gap in health access and improving health outcomes for women.

Mobile phones serve as a potent tool for healthcare across the dimensions of prevention, treatment, and cure. In collective societies like Kenya, characterized by high levels of poverty, digital health information geared toward individual autonomy may not adequately address the health needs and conditions of lower economic classes or rural women [10]. The primary focus areas of mobile phone technology interventions in developing countries, particularly in mHealth, include supporting health workers in service delivery, delivering information directly to consumers, and facilitating behavior change in disease management [10].

The widespread ownership of mobile phones in Kenya offers opportunities to develop easily accessible mobile-based knowledge aimed at influencing attitudes toward health and healthcare-seeking behaviors, particularly among women. With the extensive adoption of mobile phone technology among women in Kenya, utilizing short text messages (SMS) can serve as an effective tool for disseminating reproductive health information, thereby enhancing health outcomes. By delivering prenatal and postnatal guidance to mothers, SMS can promote preventive care and provide information regarding future appointments, consequently reducing instances of morbidity. However, the accessibility of SMS-based knowledge hinges on users' decision-making abilities and available resources [24]. Ngaruiya et al. [24] discovered that SMS interventions offer an inclusive approach to health interventions, particularly benefiting marginalized populations with limited access to healthcare. Individuals with lower levels of education and income residing in slum settlements such as Korogocho and Viwandani in Nairobi were more likely to receive and utilize SMS messages, indicating that mobile technology can play a positive role in bridging the gap in health information equity [24].

In addressing the knowledge gap between low-income and rural women regarding their rights and benefits within the National Hospital Insurance Fund (NHIF) system, mHealth can play a crucial role. Lack of awareness about NHIF benefits discourages membership and limits access to quality care. Many women are unaware of the benefits available to them within the social health system, undermining their ability to make informed health decisions, control their health-seeking behavior, and trust the system. Consequently, this lack of awareness contributes to low renewal or enrollment rates in the NHIF health scheme [19]. MHealth, through SMS messages, can serve as a powerful tool to reinforce the significance of early diagnosis for individual health outcomes.

3.5 Social Barriers to Care: Perception of Care and Resolutions

The quality of care and health outcomes are adversely affected by different forms of bias, often disguised as unprofessional conduct. Numerous studies confirm the prevalence of negative interactions experienced by individuals in lower economic strata within public health facilities, including encounters marked by hostility, rudeness, and cultural insensitivity [21]. In the Kisii community of Kenya, healthcare providers were reported to exhibit unfavorable attitudes toward inexperienced pregnant teenagers, partly due to societal disapproval of premarital sex and dropping out of school, thus compromising the care they receive [6]. Staff attitudes toward patients during maternal care were influenced by factors such as age, maternal experience, social class, and sociocultural considerations, contributing to instances of disrespectful behavior [6].

The quality of the emotional connection between healthcare providers and patients significantly influences women's health-seeking behavior and their subsequent access to quality healthcare. According to a study by Suchman et al. [19], Kenyan individuals cited the positive reputation of a facility, the quality of services offered, and the reputation of healthcare providers as determining factors in their choice of specific health facilities. However, there is a notable gap in professional and cultural competency among healthcare providers in low-resourced rural and urban areas of Kenya. Many healthcare providers prioritize their interests over those of their patients, lacking essential skills such as conscientiousness, respect, emotional intelligence, and appropriateness in their interactions. Consequently, patients often encounter unprofessional behavior, including breaches of patient-provider confidentiality, which are crucial for fostering patient trust and confidence in receiving care [15]. In a study conducted in North Eastern Kenya, women seeking pre- and post-natal services expressed dissatisfaction with the devaluing treatment they received from healthcare providers, which bordered on overt disrespect [17]. This mistreatment reportedly discouraged women from seeking future healthcare services in public hospitals. However, contrasting these findings, research by Reddy et al. [25] suggested that the health-seeking behavior of women in rural areas depended more on the cost and location of the health facility rather than the qualifications of the healthcare provider or the quality of care provided.

A study conducted by UNFPA Kenya revealed that insufficient human resources and substandard quality of care detrimentally affect women's health-seeking behavior in the country [26]. Numerous public health facilities lack multidisciplinary teams essential for delivering comprehensive care. Consequently, women encounter challenges in accessing the required healthcare services, as it remains unclear which facilities offer specialist services, diagnostic equipment, or access to current medical knowledge through information technology. The quest for appropriate treatment often leads to delays as women expend significant time and effort moving between health facilities seeking assistance. This experience fosters a negative perception of healthcare services among women. Table 1 encapsulates the

Table 1 Social barriers to care matrix

Care needs			
Social barriers			
Mobile phone technology	*Language and literacy levels*	*Social capital*	*Perception of care*
– SMS information in isolation of supply side in health behavior change – Inadequate use of mobile phones to message health information – Inadequate use of mobile phones for accessing physicians	– Low health literacy levels – Language biases – Different interaction language with health providers – Lack of awareness and knowledge of rights and benefits within the social care system – Poor individual decision-making capacity	– Rural women bonding groups instead of bridging groups that disseminate information that values and improves health	– Low priority and value of self-health care – Low appreciation of the importance of good health – Beliefs and trust in the competency of traditional healers over modern medicine – Poor treatment by health personnel – Unreliable services due to information gaps – Untrustworthy due to lack of multidisciplinary teams in health facilities for comprehensive care – Poor management and referral system – Not patient-centered
Preventative, diagnosis, treatment	*Preventative, diagnosis, treatment, cure*	*Preventative*	*Preventative. Diagnosis, treatment, cure*

(continued)

Table 1 (continued)

Resolutions (Fit between the intervention and implementation context) [5]			
– Mobile phone information messaging increases awareness, confidence, and autonomy in women decision-making – Increase preventative and experiential health information through mobile phones – Use local languages in messaging – Access to mobile-based physician dial-in service – Use SMS for health behavioral change information alongside access to accompanying supply-side health resources	– Increased SMS messaging on the rights and benefits of women in the social health system – Patient training and education on the detrimental effects of language bias	– Membership in women bridging groups for attitude and health behavior change	– Mobile phone-based information on the value of healthcare – Dedicated days for multidisciplinary team care – Sanctions for poor treatment by health personnel – SMS information on risks of delayed treatment – Record keeping on referrals – User-centered design and participatory approaches – Training and education in cultural competency

four dimensions of social barriers and corresponding resolutions we have been analyzing across four sections.

3.6 Cultural Barriers and Resolutions

Attitudinal and behavioral obstacles affecting women's healthcare experiences vary based on the degree of religious adherence and ethnic background. In numerous rural areas of Kenya, conservative societies, predominantly adhering to a single culture or religious doctrine, adhere to a communal, predictable, and conservative course in influencing women's healthcare-seeking conduct, as dictated by local religious and cultural standards [15]. The Somali community in Kenya exhibits a prevalent cultural and religious identity intertwined with patriarchal customs and belief systems that influence the healthcare encounters of Somali women. The economic and social authority of men within Somali communities constrains women's ability to resist the practices of Female Genital Mutilation/Cutting (FGM/C) [27]. This male authority and influence are substantial within Somali communities, compounded by women's strong emphasis on maintaining continuous familial connections and unquestioned social acceptance [27]. The cultural practice of FGM/C is deeply ingrained in cultural gender dynamics, social acceptance, and religious convictions. Sexual and reproductive healthcare presents particular challenges due to its association with cultural gender dynamics and unequal power structures [27].

Low-income Somali women face increased reluctance to seek sexual health care due to barriers rooted in religious and cultural beliefs, which perceive such care as offensive [15]. Those who endure complications related to FGM/C express disapproval and hesitance towards seeking medical assistance in facilities where the majority of healthcare providers are male or from different ethnic backgrounds, as this contradicts the perceived value of the practice in enhancing social acceptance and identity [15]. Cultural and religious beliefs play a significant role in shaping the healthcare-seeking patterns of pregnant women seeking prenatal and postnatal care [14]. Within reproductive health services, the Somali community in Kenya strongly discourages physical contact between female patients and male health workers, reflecting an attitude deeply rooted in the Islamic faith and Somali cultural norms [14]. Similarly, Muslim or conservative/traditional women in the Kisii and Kilifi regions of Kenya exhibit comparable reservations towards male health providers [6]. In a study in Nairobi city on the influence of sociocultural factors on access to health among various economic classes, Chelogoi et al. [21] found significant commonality among women in giving meaning to death, illness, and health through a cultural lens. Within the Kisii and Kilifi communities, entrenched cultural beliefs extend beyond gender power dynamics to encompass reproductive practices. Rural women within these societies are stigmatized with the notion of experiencing ominous childbirths if deliveries do not occur within specific timeframes, leading them to resort to herbal remedies to expedite the birthing process. Women within these communities exhibit a strong inclination towards and trust in home births, drawing from extensive personal experiences, which in turn diminishes their receptiveness to modern medical practices [6].

In response to the cultural challenge of gender sensitivity in health care provision, the county government of Garissa prioritized the hiring of female health workers in reproductive health over male attendants [14]. This policy has opened another layer of challenges to quality healthcare provision, by putting at risk the quality of healthcare received due to understaffed and overworked female health workers. The health care management system in Kenya is highly decentralized with no structures facilitating coordination and administrative and clinical integration. There is very little communication or collaboration between public health facilities and slow adoption of information technology innovations needed to provide care. This anomaly affects the quality of care and increases the cost of care due to repeated diagnostic treatment and unnecessary health provider visits.

New health providers are unaware of the medical backgrounds of patients, resulting in fragmentation of care and conflicting or overlapping treatment which could result in adverse health outcomes or patient death. This may generate frustration and delay treatment of a potentially adverse condition affecting patient trust levels [14]. Table 2 describes the three dimensions of cultural barriers to healthcare and the corresponding resolutions we have been analyzing in this section.

Within the social care system, the lack of accessible information regarding facility accreditation status restricts the options available to women seeking quality services at lower costs. However, leveraging a mobile health (mHealth) pathway could effectively disseminate this information to women in low economic status groups or residing in rural areas, thereby broadening their access to quality care [28]. Currently, no established health management system directs women with conditions such as complications from FGM/C to facilities equipped to address them. Consequently, women are compelled to self-refer, potentially leading to subpar treatment in facilities lacking medical specialists trained to manage FGM/C-related complications. This increases the likelihood of medical errors with potentially fatal health repercussions and may deter women from seeking timely and appropriate treatment for FGM/C-related issues [15].

4 Conclusion

The goal of this chapter was to highlight literature on the socioeconomic and cultural barriers to healthcare using the lens of personhood to deepen understanding of cultural and religious forces on women's health decision-making and the consequences on health equity [29]. As described in Fig. 2, a conceptual framework on the relationship between women's health decisions and health inequities in Kenya has been designed to deepen understanding and facilitate intervention formulation. This chapter finds disparities exist largely for vulnerable women groups in low economic brackets in rural and urban areas. The cultural barriers go beyond contextual cultural influences and include implicit bias based on language use and socio-economic status as well as education levels. The meeting of differential economic status, varied education levels, and language differences may hurt health providers' motivation to give quality care.

Table 2 Cultural barriers to the care matrix

Care needs		
Cultural barriers		
Dominant cultural beliefs and practices	Dominant religious beliefs and practices	Patriarchy
– Women value of relational connectedness – Women hold in high esteem social approval and acceptance over individual rights and freedom – Value of appearing marriageable – Social stigma – Social discrimination – Ethnicity and gender of health providers – Cultural myths – Preference for traditional home birth deliveries	– Limits contact between males and females – Practices seen as religiously offensive	– Male authority and dominance – Males have the final decision on women's healthcare – Women value their cultural representation of attractiveness and approval by men – Man's authority over women's attractiveness standards – Female lack of decision-making power – Lack of women's access and control over resources – Inequity in gender resource allocation
Preventative, diagnostic, treatment, and cure	Preventative, diagnostic, treatment, and cure	Preventative, diagnostic, treatment, and cure
Resolutions		
– Train more female health workers in sexual and reproductive health – Public meetings on the health risks and illegality of practices like FGM/C – Mobile phone messaging to women on FGM/C awareness and knowledge	– Train more female health workers in sexual and reproductive health – Digital messaging on the importance of healthcare – Cultural sensitivity training and education for health providers	– Community sensitization on the value of health – Involvement of local women empowerment advocacy groups

To improve health outcomes for women and narrow the equity gap, interventions need to take into account, the meaning of personhood within local communities and integrate that with knowledge of communities, women, health facility personnel, and other realities such as financial position and religious value to arrive at a holistic understanding of the interactions of all these forces on women care and continued health inequity. Resolutions will then be based on local realities and understanding.

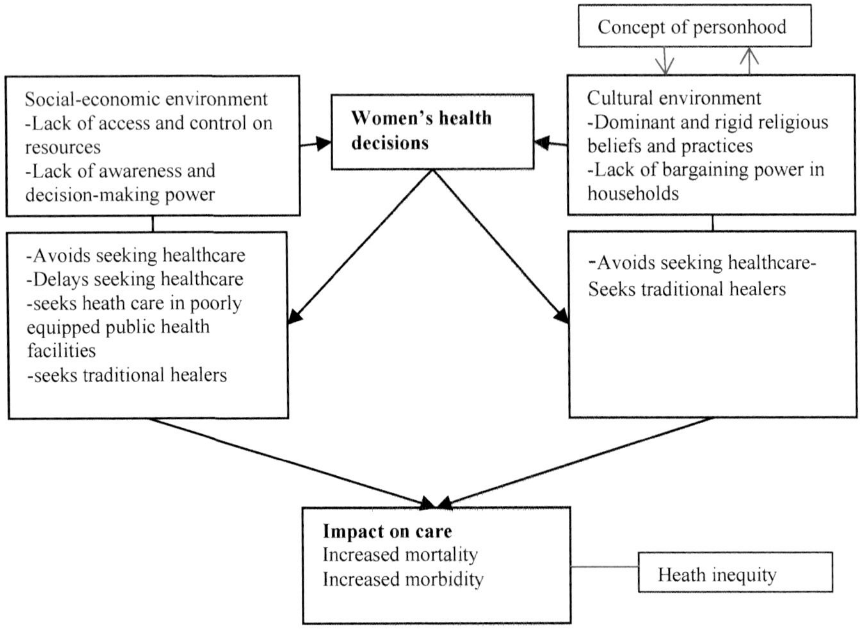

Fig. 2 Conceptual framework on the relationship between women's health decisions and health inequities in Kenya

References

1. USAID (2022) Kenya: gender equality and female empowerment factsheet. https://www.usaid.gov/sites/default/files/2022-05/Gender_Fact_Sheet_for_Kenya.pdf. Accessed 6 Jan 2024
2. Osok J, Kigamwa P, Huang KY, Grote N, Kumar M (2018) Adversities and mental health needs of pregnant adolescents in Kenya: identifying interpersonal, practical, and cultural barriers to care. BMC Womens Health 18(1):1–18
3. Alushula P (2023). Business Daily. https://www.businessdailyafrica.com/bd/economy/kenyans-to-pay-2-75-pc-of-monthly-earnings-to-fund-healthcare--4448790. Accessed 6 Jan 2024
4. Artiga S, Hinton E (2018) Beyond health care: the role of social determinants in promoting health and health equity. Kaiser Family Foundation 10
5. Baumann AA, Cabassa LJ (2020) Reframing implementation science to address inequities in healthcare delivery. BMC Health Serv Res 20:1–9
6. Lusambili A, Wisofschi S, Shumba C, Obure J, Mulama K, Nyaga L et al (2020) Health care workers' perspectives of the influences of disrespectful maternity care in rural Kenya. Int J Environ Res Public Health 17(21):8218
7. Kenya National Bureau of Statistics (2022) Kenya demographic and health survey. Kenya National Bureau of Statistics, Nairobi
8. Johnson TJ (2020) Intersection of bias, structural racism, and social determinants with health care inequities. Pediatrics 146(2):e2020003657. https://doi.org/10.1542/peds.2020-003657
9. Kpanake L (2018) Cultural concepts of the person and mental health in Africa. Transcult Psychiatry 55(2):198–218

10. McCool J, Dobson R, Whittaker R, Paton C (2022) Mobile health (mHealth) in low-and middle-income countries. Annu Rev Public Health 43:525–539
11. Kenya National Bureau of Statistics (2019) Kenya population and housing census. https://www.knbs.or.ke/2019-kenya-population-and-housing-census-results. Accessed 10 Jan 2024
12. Jennings L, Gagliardi L (2013) Influence of mHealth interventions on gender relations in developing countries: a systematic literature review. Int J Equity Health 12(1):1–10
13. Ministry of Public Service, Youth and Gender Affairs (2017) Kenya Country Report. Women's Economic Empowerment in the Changing World of Work. https://gender.go.ke/wp-content/uploads/2022/06/CSW-61st-Kenya-Country-Report.pdf. Accessed 21 Nov 2023
14. Kisiangani I, Elmi M, Bakibinga P, Mohamed SF, Kisia L, Kibe PM, Otieno P et al (2020) Persistent barriers to the use of maternal, newborn and child health services in Garissa sub-county, Kenya: a qualitative study. BMC Pregnancy Childbirth 20:1–12
15. Kimani S, Kabiru CW, Muteshi J, Guyo J (2020) Exploring barriers to seeking health care among Kenyan Somali women with female genital mutilation: a qualitative study. BMC Int Health Hum Rights 20:1–12
16. The National Treasury (2023) The Republic of Kenya Health Sector Report 2022/2023. https://www.treasury.go.ke/wp-content/uploads/2023/12/HEALTH-SECTOR-REPORT.pdf. Accessed 6 Jan 2024
17. N'Gbichi C, Ziraba AK, Wambui DW, Bakibinga P, Kisiangani I, Njoroge P et al (2019) If there are no female nurses to attend to me, I will just go and deliver at home: a qualitative study in Garissa, Kenya. BMC Pregnancy Childbirth 19:1–9
18. Knoema (2021) World Data Atlas Kenya Health. https://knoema.com/atlas/Kenya/Neonatal-mortality-rate. Accessed 4 Dec 2023
19. Suchman L, Hashim CV, Adu J, Mwachandi R (2020) Seeking care in the context of social health insurance in Kenya and Ghana. BMC Public Health 20(1):1–13
20. Musalia J (2016) Social capital and health in Kenya: a multilevel analysis. Soc Sci Med 167:11–19
21. Chelogoi DN, Jonyo FO, Amadi H (2020) The influence of socio-cultural factors in access to healthcare in Kenya: a case of Nairobi County, Kenya. Open J Soc Sci 8(5). https://doi.org/10.4236/jss.2020.85023
22. Communications Authority of Kenya (2023) Mobile subscriptions. https://www.ca.go.ke/mobile-subscriptions-hit-66m-march-2023. Accessed 1 Dec 2023
23. WorldData.info (2022) https://www.worlddata.info/africa/kenya/telecommunication.php. Accessed 10 Dec 2023
24. Ngaruiya C, Oti S, Van de Vijver S, Kyobutungi C, Free C (2019) Target women: equity in access to mHealth technology in a non-communicable disease care intervention in Kenya. PLoS ONE 14(9):e0220834
25. Reddy PMC, Rineetha T, Sreeharshika D, Jothula KY (2020) Health care seeking behaviour among rural women in Telangana: a cross sectional study. J Family Med Primary Care 9(9):4778
26. UNFPA Kenya (2022) UNFPA, Kenya Annual report 2021. https://kenya.unfpa.org/en/publications/unfpa-kenya-annual-report-2021. Accessed 29 Nov 2023
27. Abathun AD, Sundby J, Gele AA (2016) Attitude toward female genital mutilation among Somali and Harari people, Eastern Ethiopia. Int J Women's Health 8:557–569. https://doi.org/10.2147/IJWH.S112226
28. Ampt FH, Lim MS, Agius PA, L'Engle K, Manguro G, Gichuki C et al (2020) Effect of a mobile phone intervention for female sex workers on unintended pregnancy in Kenya (WHISPER or SHOUT): a cluster-randomised controlled trial. Lancet Glob Health 8(12):e1534–e1545
29. Claridge T (2018) Functions of social capital–bonding, bridging, linking. Soc Capital Res 20(1):1–7

Addressing Gender Health Inequity: Bridging the Gap in Sexual and Reproductive Health Rights for Young Girls

12

Ni Komang Yuni Rahyani

There is much to be done to fulfill global unmet adolescent contraceptive needs. Adolescent continuous to have difficulty in accessing and using modern contraceptive methods due to significant stigma and bias around contraception use

Dr. Aparna Sridar. Chair, FIGO Committee on Contraception

Abstract

Adolescent girls are an age group that is very vulnerable to injustice in obtaining their reproductive rights. Marital status determines the type of contraceptive services that can be provided to women. Sexually active adolescent girls will receive a negative stigma when requesting contraceptive services at health facilities. The providers' attitudes tended to be judgmental and considered teenage girls to be naughty children. There is a perception that if adolescent girls are provided with contraceptive services, it appears as if health workers support premarital sexual behavior. The main problem related to injustice in fulfilling the reproductive rights of adolescent girls includes unsafe abortion due to an unmet need for birth control services. Most unsafe abortions are triggered by unwanted pregnancies due to limited access to contraceptive services for adolescent girls. Lack of knowledge, economic dependence on parents, as well as different policies in each area regarding sexual and reproductive health services for adolescent girls. Another problem is intimate partner violence (IPV), which has a serious impact on the health and well-being of adolescent girls

N. K. Y. Rahyani (✉)
Department of Midwifery, Ministry of Health, Polytechnic of Health of Denpasar, Denpasar, Indonesia
e-mail: rahyaniyuni1@gmail.com

B. W. K. Son (ed.), *The Landscape of Global Health Inequity*, Integrated Science 22,
https://doi.org/10.1007/978-3-031-60502-4_12

in the future. Efforts that can be offered are to review policies related to reproductive and sexual health services for adolescents through prevention and promotion efforts. Cross-program and cross-sector collaboration is needed to obtain optimal results.

Keywords

Adolescence • Gender health inequity • Prevention • Collaboration • Family planning programs

1 Introduction

Fairness in providing reproductive and sexual health services for the adolescent age group, especially adolescent girls, is still a matter of debate. The Center for Reproductive Rights stated that various indicators of disparities in reproductive and sexual health services for women include: limited access to comprehensive reproductive and sexual health education regarding types of contraception, and efforts to prevent unwanted pregnancies including unsafe abortions. The fact previously disclosed is that one in four sexually active teenagers do not receive the contraceptive services they should receive. This condition increases the incidence of unwanted pregnancies among young girls, and around seven million adolescent girls give birth to babies every year [1]. According to the United Nations Population Fund (UNFPA) in FIGO, the highest incidence of unwanted pregnancy occurs among the adolescent age group 15–19 years, where 57% is caused by an unmet need for contraception [2, 3].

The World Health Organization (WHO) has issued six main recommendations since 2011 to fulfill adolescents' rights regarding reproductive and sexual health. The six recommendations are in the form of: (1) reducing the number of marriages among teenagers (aged < 18 years); (2) reducing the number of pregnancies in women aged < 20 years; (3) increasing the use of contraception by teenagers who are at risk of experiencing unexpected pregnancy; (4) reducing the incidence of rape against teenage girls; (5) reducing the incidence of unsafe abortion; and (6) improving the quality of care for adolescent girls during pregnancy, childbirth and postpartum [4]. However, these recommendations have not been widely accepted, especially in developing countries. Efforts to use contraception among at-risk adolescent girls and safe management of abortion are very limited and even very difficult to obtain. Studies in Kenya found that the highest mortality among women aged 15–19 years was due to unsafe abortion, where teenage girls did not use contraception before pregnancy occurred [5]. Based on a systematic review, it was found that the knowledge, attitudes, and actions of adolescents regarding teenage pregnancy and unsafe abortion were correlated with low access for unmarried adolescents to the utilization of contraceptives and safe abortion services in Sub-Saharan [6].

Adolescent girls experience more negative consequences including discrimination and stigma by society, teachers at school, and their peers when experiencing pregnancy before marriage and complications due to unsafe abortion [5, 7]. The occurrence of pregnancy and unsafe abortion in unmarried teenage girls is considered an appropriate punishment due to behavior that violates norms by society, including by health workers. The results of a study by Sully et al. [2], found that an estimated 21 million teenagers aged between 15 and 19 years experienced pregnancy. Approximately 50% of these pregnancies are unintended, and 55% of unintended pregnancies end in unsafe abortions. This condition generally occurs in adolescent girls in low- and middle-income countries [8].

The role of parents, education by teachers at school, the influence of the mass media, and health workers are very important as sources of appropriate information about reproductive and sexual health for adolescents. On the other hand, this role has not been implemented due to various obstacles. limited knowledge of parents, considering it taboo, feeling embarrassed, low competence of health workers regarding efforts to promote reproductive and sexual health in health service facilities, limited access of teenagers to health facilities when they need contraception, economic dependence, as well as feelings of shame or fear of being blamed are influencing factors. regarding the non-fulfillment of adolescent rights related to reproductive and sexual health [3, 7, 9].

To face challenges related to fulfilling reproductive and sexual health rights in adolescents, recommendations need to be prepared that directly provide effective and efficient solutions and can be widely applied. The unmet need for contraception covers a very complicated and complex part, due to the concerns of global and national policymakers regarding controlling population growth and fulfilling reproductive and sexual rights for women [10, 11].

2 Gaps in the Quality of Reproductive and Sexual Health Services for Adolescent Girls

In general, factors related to the use of modern contraception by adolescent girls are (1) socio-cultural and socio-demographic factors including cultural beliefs; (2) peer influence; (3) religion; (4) as well as contraceptive side effects [12, 13]. The results of other studies found that the factors of affordability of contraceptive services, priority level of need, level of education, socio-economics, women's autonomy, sex education, and competence of health providers influence contraceptive use [14]. According to the United Nations Population Fund (UNFPA), some pregnancies in teenage girls are unexpected pregnancies. This condition is a form of crisis in the completion of human rights. Women, wherever they are and at any time, are expected to have the ability to make choices and to have power over their bodies [3]. Pregnancy in a married woman is an expected event for most couples. However, there are different conditions if pregnancy occurs in teenage girls, especially those who are not married. Pregnancy examination services are prioritized for married couples, including funding from health insurance. According to the

World Health Organization (WHO), the number of visits for antenatal care during pregnancy to health care facilities is at least six times [15]. In 2016, they released the implementation of routine antenatal care for pregnant women and adolescent girls. These recommendations were prepared with the aim of improving the health and welfare of women and their babies [16].

The patriarchal culture that is still practiced in various countries in the world requires women to maintain their virginity until marriage. The society also has attitudes and values that agree that teenage girls who have sexual relations outside of marriage are generally perceived as naughty, wild, or as having low morals [6, 9, 17–19]. The results of previous studies in Africa showed that there was a stigma towards sexually active adolescent girls. This stigma takes the form of being labeled as a naughty child, immoral, embarrassing, and other forms of discrimination [20–22]. Stigma, labeling, and moral sanctions by society against women and girls who are sexually active outside of marriage are factors that trigger the unmet need for modern contraception. Another factor is financial dependence on parents in addition to actions from health and medical services that require parental consent [5, 7, 11, 18, 19, 23–26].

Studies on the determinants of contraceptive use among adolescents are determined by marital status, religion [9], knowledge, sources of information about contraception [6], age, occupation, socio-economic status, and geography [7]. A similar study in Kenya found that the prevalence of unmet needs in adolescents and girls aged between 15 and 19 years was 28.3% (95% CI 25.7–31.0). Various contributing factors include low socioeconomic level (aOR: 5.6, 95%CI: 2.8, 11.1), never being exposed to the media (aOR: 2.1, 95%CI: 1.1–4.1), and living in a developing region (aOR: 5.1, 95%CI: 1.1–14.5) [18]. Another study in developing countries such as Indonesia found that the incidence of unmet needs in Indonesia among women aged 15–49 years was 14.3% (95% CI 13.9–14.7%). The determinants of unmet needs in Indonesia are the woman's age, age of last child > 10 years, having > 1 son, husband not working, low knowledge about contraception, living in an urban area, and religion [27]. Based on these conditions, unmet need is related to sociodemographic conditions, low levels of education and knowledge, low socio-economic levels, limited access to health facilities, low competence of health providers, including religious values believed in [7, 9, 17–19, 25–27].

Various determinants of the unmet need for contraception in adolescents have been described using a social-ecological theory approach developed by Bronfenbrenner [28] and adopted by Inanc et al. [24]. Based on the social-ecological theory, three factors were obtained that influence the unmet need for contraception, including: (1) Individual factors, namely the characteristics of adolescent girls; (2) Interpersonal factors including family factors, peer influence, and support from intimate partners; and (3) Environmental factors in the form of the influence of policies, programs, media exposure and health systems and others [24]. Individual factors in the form of characteristics of adolescent girls are both risk factors and protective factors. Risk factors include premarital sexual behavior and other risky behavior in the form of biological factors (gender/sex, earlier age at menarche), and low knowledge factors. Protective factors include healthy behavior factors (not smoking, not drinking alcohol, or using illegal substances), positive intention and belief factors, psychological well-being factors, and skills. The most influential interpersonal factors as protective factors are teenagers who live with complete parents, are highly educated, have high socio-economic status, have closeness, and have regular and open communication with their parents. Environmental factors that have a positive influence on adolescent behavior are the characteristics of the living environment, especially neighbors who care about each other, the influence of positive media and mass media exposure, and the service system, including the health system [13, 24, 29, 30].

Determinants and risk factors related to unmet need for contraceptive services variations are very wide in each country and will influence the policies and programs that will be selected. Health policymakers need to understand the risks of ignoring the fulfillment of the reproductive and sexual health rights of women and adolescent girls. The risk of maternal death is four times higher in adolescent girls who give birth at less than 16 years of age compared to women aged > =20 years [31]. The negative impact of pregnancy and childbirth on adolescent girls whose body growth is not yet complete causes obstetric fistulas which generally occur in adult women at 9–86% [32]. Pregnancy among adolescent girls is more likely to give birth to premature babies, low birth weight babies, and a lack of ability to care for babies properly [4, 31, 33]. Another negative impact due to pregnancy and childbirth at a very young age is mental problems in the form of depression [34].

3 Strategy for Fulfilling Adolescent Rights Regarding Reproductive and Sexual Health Services

An indicator that shows that fertility is still high in Sub-Saharan Africa is limited access to contraceptive services [25]. Similar conditions also occur in the Guinea region, especially in the group of female adolescents aged between 15 and 19 years [19]. The incidence of unmet needs in India is around 12.8% experienced by married women [35]. The study in India used secondary data, namely the National Family Health Survey, regarding the determinants of unmet need for contraception in the form of religion, age, residence in rural areas, youth, and living in poverty. The incidence of unmet needs among Muslim women in India is 32%, where they live in rural areas (24%), in the teenage age group (36%), and are classified as poor (26%) [26].

The results of a previous survey by Guttmacher through the Demographic Health Survey (DHS) survey between 1995 and 2005 showed that one of the causes of unmet needs in women was low knowledge about contraception. However, survey results in the 1980s showed that the factor of low knowledge about contraception was higher when compared with survey results after the 1995 period [11]. The cause of unmet needs in women of childbearing age is that they do not have sexual intercourse regularly and are still breastfeeding their babies. Thus, women perceive that their chances of becoming pregnant are low [10].

The unmarried adolescent age group experiences more obstacles in fulfilling their needs related to the use of contraception when compared to the married female group. Cultural factors, religion, and values believed in society are factors inhibiting the fulfillment of contraceptive services for unmarried women [7, 19, 36]. Evidence shows inequality in fulfilling the right to obtain modern contraceptive services for sexually active adolescent girls. Studies in Kenya found that the incidence of unwanted pregnancy in teenage girls aged between 15 and 19 years was the reason why teenage girls had unsafe abortions [5], and had fatal consequences, namely death due to post-abortion bleeding or infection. The paradigm from teachers at school, peers, and peer counselors regarding the use of contraception in teenage girls is that teenagers are naughty, wild, and set a bad example for other friends. Another study in Sub-Saharan Africa (SSA), found that the highest prevalence of unmet needs was in Angola at 24.9% and the lowest in Nigeria (17.8%). Young women aged between 20 and 24 years, living in rural areas, and having a lower level of education are indicators of higher unmet need [7, 21, 25, 37, 38]. In contrast, a study in Ghana found that adolescents aged between 15 and 19 years were sexually active and did not use contraception during sexual intercourse, more than 50% had high knowledge regarding contraception [12]. Injectable contraception is one of the types of contraception most often chosen by adolescent girls based on considerations from health workers (12).

From the various evidence above, various obstacles have been obtained or there has been no realization of a global agreement that gives priority to adolescent girls regarding fulfilling the right to information and quality contraceptive services, especially in low- and middle-income countries. Next, we will describe the policies that have been prepared globally and nationally as a comparison in developed and developing countries. For example, WHO has issued recommendations since 2016 regarding pregnancy as a pleasant experience [16]. Not only that, WHO has also issued a strategy to reduce morbidity and death rates for women due to pregnancy and childbirth in the form of Survive, Thrive and Transform [39]. Pregnant women should have at least eight contacts with trained health personnel, especially doctors, obstetricians, midwives, and nurses, at a minimum of eight times in health facilities. However, in Indonesia, this policy was only implemented around 2018 with a minimum number of antenatal care visits being six times [40]. What about strategies and policies for fulfilling rights regarding safe contraceptive information services for sexually active teenagers? From evidence in several countries, it was found that there are still obstacles and difficulties in providing modern contraceptive services, education, and counseling services for teenagers, including safe abortion services due to unwanted pregnancy.

Based on data from the United Nations Department of Economic and Social Affairs, Population Division (2020), it was found that the teenage birth rate has decreased from 60 to 43 births per 1000 girls aged 15–19 years over the last 25 years. The data was collected from 1995 to 2000 and from 2015 to 2020. The decline in teenage birth rates in South Asia has made the greatest progress, alongside North America, Sub-Saharan Africa, and Latin America. However, globally this figure varies between countries or between regions and even within regions. For example, the teenage birth rate in South Asia in 2015–2020 was 26 per 1000 girls aged 15–19 years, while in Sub-Saharan Africa it was 104 per 1000 girls aged 15–19 years. Based on 2018 basic health research data in Indonesia, it was found that 58.8% of the population involved in the survey (N = 4861 married women aged between 15 and 19 years) admitted to having been pregnant. Furthermore, as many as 25.2% of the 2867 women admitted to being pregnant when they were involved in the survey [41].

Even though the results of various previous studies have found various negative impacts from not fulfilling the rights of adolescent girls regarding the use and access to contraception to prevent unwanted pregnancies, to date no significant changes have been found. In other words, the number of unmet needs for contraception in the adolescent age group, sexually active women, has not shown a significant decline. WHO and UNFPA have issued guidelines since 2011 regarding the prevention of teenage pregnancy and its negative impacts in developing countries [42]. The strategies or efforts that have been prepared in various countries, especially in developing countries or countries with low—middle income, are outlined in Table 1 as follows:

There are three benefits from investing in teenagers globally, including profits for teenagers now, for later life in adulthood or in the future, and for children who will be born in the future. Based on this, various obstacles faced by groups of teenagers are described in obtaining their rights, especially preventing and avoiding unwanted pregnancies through fulfilling contraceptive services. These obstacles and challenges include (1) the low ability of adolescents to prevent, avoid, delay and limit the number of pregnancies and pregnancies; (2) low desire to use contraception; (3) limitations in the ability to finance independently and dependence on parents or partners to use contraception; (4) there are obstacles to obtaining contraceptive services due to regulations regarding age and marital status; (5) related to the behavior of service providers in health facilities who are incompetent and unfriendly towards teenagers [43–45].

Based on the data from the United Nations (2022), there are differences between the needs of adolescent girls and young girls in various areas with the use of modern contraception. Differences based on age groups in meeting needs related to modern contraception are quite small. More than 75% of female contraceptive acceptors of all ages are satisfied with the modern contraceptive services they have received, especially in Latin American, Caribbean, Australian, European, and North American countries. Meanwhile, in low-income countries, it was found that the level of satisfaction with modern contraceptive services was less than 58%, and in the adolescent age group, the level of satisfaction was 41%.

Another problem faced by adolescents apart from unmet need for contraception and unsafe abortion, the problem of gender inequality has become a serious problem. Gender inequality or injustice, which is often called gender-based violence (GBV), is experienced by women and adolescent girls in various countries

Table 1 Prevention strategies for teenage pregnancy

No.	Country	Policy/strategy
1	Chile	The problem is related to the high fertility rate in 2005, namely 55.8 births per 1000 women aged between 15 and 19 years. Adopt the regional 2007–2013 Andean Plan for the Prevention of Adolescent Pregnancy. The target to be achieved is to reduce the fertility rate by 10% from 2011 to 2020 [43]. The involvement of various sectors that support the achievement of these targets includes advocacy with advocates, non-governmental organizations, non-governmental organizations (NGO's), primary health care centers, schools, women's activists and other sectors who want to fight for the rights of adolescent girls to obtain education and contraceptive services An approach through five pronged efforts aimed at increasing sensitivity and awareness in the health services sector for the next tenyear strategy, including: a. train health workers, b. creating youth-friendly spaces in primary health centers, c. promote various contraceptive methods, d. increasing reach and referrals, and e. increasing school retention and re-entry for pregnant teens and teens who are becoming mothers
2	Ethiopia	Negative stigma is attached to adolescent girls who need contraception to prevent unwanted pregnancies and prevent unsafe abortions due to unmet need for contraception (6). Increasing coverage of contraceptive use in Ethiopia is aimed at reducing maternal mortality. The Health Extension Program (HEP) program has been launched containing a health education and agenda basic health services including a basket of contraceptive methods in the community, and to strengthen links to health facilities. However, the program is directed at rural areas because more than 80% of the population lives in rural areas. The program also targets groups of teenagers who are sexually active even though they are not married The government recruits and trains cadres to help health workers provide education and information to the public, change norms and increase the use of health services related to contraception The government also involves various associations, especially health professional associations, educational institutions and NGOs The results of the policies that have been developed and implemented in Ethiopia are that there has been an increase in the use of modern contraception among married women aged 15–19 years from 2000 to 2016 (from 3 to 31.8%). Likewise, the use of modern contraception among women in the postpartum period was 38% in the same age group [43]
3	Nigeria	According to data from DHS in 2013 in Nigeria, it was found that more than 30% of mortality among adolescent girls aged between 15 and 19 years were associated with pregnancy and the birth of a baby. The program approach to reducing unmet need in Northern Nigeria among adolescent girls is through efforts that target men, involve religious leaders, and overcome religious and socio-cultural norms that hinder the use of contraception (43) Limited information on how to prevent unwanted pregnancies in teenagers, limited access to contraceptive services, the attitude of health workers who often blame teenagers who are sexually active outside of marriage, and a culture that considers it taboo to discuss issues of sexuality and reproductive health as the main causes of high unmet need. for contraception in adolescent girls aged 15–19 years, especially in Africa [6]

in the world. Part of GBV is intimate partner violence (IPV) which equally harms women's reproductive and sexual health in the future [46]. Intimate partner violence was found to be higher in women as victims and men as perpetrators. The incidence of IPV increases a woman's risk of experiencing negative outcomes ranging from mild to severe. These negative outcomes include physical and mental health problems in victims [47]. Based on the problems or disparities related to incidents of unmet need for contraception, gender-based violence (GBV), and IPV, more of the victims are women, especially adolescent women. Therefore, it is necessary to consider in-depth strategies that are more focused on improving the quality of services, including the affordability of health and reproductive services for adolescent women.

Sexual and reproductive health services generally cover nine areas, namely: (1) contraception/pregnancy prevention; (2) HIV/AIDS; (3) sexually transmitted infections (STIs); (4) abortion; (5) antenatal care/delivery/postnatal care; (6) reproductive tract infections; (7) infertility; (8) sexuality; and (9) partner violence (including against violence against women) and related practices (e.g., female genital mutilation) [3, 4, 46]. The field of reproductive health currently prioritizes women who are married or in a marriage relationship compared to providing services to the adolescent age group who are not or are not married. This inequality is based on several factors, especially physical and non-physical factors. Physical factors include the availability of infrastructure, equipment, competent service providers, as well as available educational and promotional media. Non-physical factors include funding by the government regarding services, tools and facilities, policies and programs that support comprehensive reproductive and sexual health services, culture, norms, religion, level of knowledge, and socio-demographic characteristics of the community [18, 48–50].

4 Conclusion and Recommendations

The main challenge faced by planners and policymakers in the field of reproductive and sexual health is that there is still a gap in efforts to fulfill the rights of women, especially adolescent girls, regarding contraceptive services. The still high unmet need for contraception among adolescent girls is caused by low levels of education, and socio-economic levels, in addition to the low quality of contraceptive service providers. The stigma attached to sexually active teenage girls who need contraceptive services has an impact on the low level of fulfillment of the need for contraception. Another problem is intimate partner violence which is still a global problem. A gap exists between low-middle income countries regarding satisfaction with the results of modern contraceptive acceptor services in each age group.

Contraceptive services by health providers will differ for potential acceptors depending on marital status. Adolescent girls who need contraceptive services will be labelled as naughty, wild, violating norms, and can have a bad influence on other adolescent girls. If health providers provide contraceptive services for adolescent girls, they will usually feel they support premarital sexual behavior. Integrated efforts are needed that involve cross-sectors and cross-programs to reduce unmet needs for adolescent girls, especially those who are not yet married, through sustainable preventive and promotional efforts.

Recommendations that can be given as launched by UNFPA are UNFPA's visionary new family planning strategy: Expanding Choices Ensuring Rights in a Diverse and Changing World, namely in the form of an approach to the future to meet contraceptive needs and prevent unmet needs in different situations and settings and anticipate current issues and trends [1–3, 51]. Strategic efforts to bridge the gap in reproductive and sexual health services for women and adolescent girls are carried out by involving cross-sector and cross-program participation. Integrated and continuous efforts are needed regarding programs, policies, services, human resources, and infrastructure including financing [43].

UNFPA and partners have recommended eight action plans for family planning programs for women and adolescent girls. These actions consist of (1) expanding affordability and access to contraceptive services; (2) Involving teenagers and young people as agents of change and providing evidence of gender-based service responses; (3) strengthening data by disaggregating data as material for program evaluation as well as financing and obstacles; (4) Increasing the sustainability of national family planning programs; (5) improve the quality of services (counseling, skills and competency of officers); (6) deepen the integration of family planning programs into national health policies, strategies, and plans including basic health services and Universal Health Coverage (UHC); (7) build resilience and adaptability; (8) able to eliminate discrimination in contraceptive services for women living in rural/peripheral areas. So, if summarized, UNFPA's actions regarding the contraception program include services that must fulfill women's human rights, a gender transformation approach, leaving no one behind and reaching the furthest behind, and considering accountability, transparency, and efficiency [3].

References

1. Center for Reproductive Rights (2023) Adolescent sexual and reproductive health and rights [Internet]. Adolescent Sexual and Reproductive Health and Rights. Available from: https://reproductiverights.org/our-issues/adolescent-sexual-and-reproductive-health-and-rights/
2. Sully EA, Biddlecom A, Darroch JE, Riley T, Ashford LS, Lince-Deroche N et al (2020) Adding it up: investing in sexual and reproductive health 2019. Guttmacher Inst [Internet], pp 1–56. Available from: https://www.guttmacher.org/report/adding-it-up-investing-in-sexual-reproductive-health-2019
3. UNFPA (2022) UNFPA strategy for family planning, 2022–2030: expanding choices—ensuring rights in a diverse and changing world. Laser Focus World 39(11):60
4. World Health Organization (2011) WHO guidelines on preventing early pregnancy and poor reproductive outcomes. Available from: http://whqlibdoc.who.int/publications/2011/9789241502214_eng.pdf?ua=1
5. Håkansson M, Super S, Oguttu M, Makenzius M (2020) Social judgments on abortion and contraceptive use: a mixed methods study among secondary school teachers and student peer-counsellors in western Kenya. BMC Public Health 20(1):1–13

6. Munakampe MN, Zulu JM, Michelo C (2018) Contraception and abortion knowledge, attitudes and practices among adolescents from low and middle-income countries: a systematic review. BMC Health Serv Res 18(1):1–13
7. Wondimagegne YA, Debelew GT, Koricha ZB (2023) Barriers to contraceptive use among secondary school adolescents in Gedeo zone, South Ethiopia: a formative qualitative study. BMJ Open 13(3):1–10
8. Sully EA, Biddlecom A, Daroch J, Riley T, Ashford L, Lince-Deroche N et al (2020) Adding it up: investing in sexual and reproductive health 2019. Guttmacher Institute, New York
9. Amoah EJ, Hinneh T, Aklie R (2023) Determinants and prevalence of modern contraceptive use among sexually active female youth in the Berekum East Municipality, Ghana. PLoS One [Internet] 18(6 June):1–18. Available from: https://doi.org/10.1371/journal.pone.0286585
10. Sedgh G, Hussain R (2007) Women with an unmet need for contraception in developing countries and their reasons for not using a method Akinrinola Bankole and Susheela Singh occasional report no. 37. Occas Rep 37:1–80
11. Sedgh G, Ashford LS, Hussain R (2016) Unmet need for contraception in developing countries: examining women's reasons for not using a method. Guttmacher Inst [Internet] 1–93. Available from: https://www.guttmacher.org/sites/default/files/report_pdf/unmet-need-for-contraception-in-developing-countries-report.pdf
12. Ziblim SD et al (2022) Sexual behaviour and contraceptive uptake among female adolescents (15–19 years): a cross-sectional study in Sagnarigu Municipality, Ghana. Ghana J Geogr 14(1):141–54
13. Salam RA, Faqqah A, Sajjad N, Lassi ZS, Das JK, Kaufman M et al (2016) Improving adolescent sexual and reproductive health: a systematic review of potential interventions. J Adolesc Heal 59(2):S11-28
14. Usha Ranji ML et al (2019) Beyond the numbers: access to reproductive health care for low-income women in five communities. Henry J Kaiser Fam Found [Internet] 2019. Available from: https://files.kff.org/attachment/Executive-Summary-Beyond-the-Numbers-Access-to-Reproductive-Health-Care-for-Low-Income-Women-in-Five-Communities
15. Tunçalp O, Pena-Rosas JP, Lawrie T, Bucagu M, Oladapo OT, Portela A et al (2017) WHO recommendations on antenatal care for a positive pregnancy experience—going beyond survival. BJOG Int J Obstet Gynaecol 124(6):860–862
16. World Health Organization (WHO) (2016) WHO recommendations on antenatal care for a positive pregnancy experience
17. Uthman MK, Bello IS, Fadugbagbe AO, Olajubu TO, Ismail WO, Ibrahim AO (2022) Unmet needs for family planning and its determinants among women of reproductive age in Ilesha Southwest Nigeria: a cross-sectional study. J Med Access 6
18. Asmamaw DB, Negash WD (2023) Unmet need for family planning and associated factors among adolescent girls and young women in Ethiopia: a multilevel analysis of Ethiopian Demographic and Health Survey. Contracept Reprod Med [Internet] 8(1):1–11. Available from: https://doi.org/10.1186/s40834-022-00211-x
19. Sidibé S, Grovogui FM, Kourouma K, Kolié D, Camara BS, Delamou A et al (2022) Unmet need for contraception and its associated factors among adolescent and young women in Guinea: a multilevel analysis of the 2018 Demographic and Health Surveys. Front Glob Women's Heal 3(Nov)
20. Hall KS, Manu A, Morhe E, Dalton VK, Challa S, Loll D et al (2018) Bad girl and unmet family planning need among Sub-Saharan African adolescents: the role of sexual and reproductive health stigma. Qual Res Med Healthc 2(1):55–64
21. Jain A, Ismail H, Tobey E, Erulkar A (2019) Stigma as a barrier to family planning use among married youth in Ethiopia. J Biosoc Sci 51(4):505–519
22. Polat F, Şenol DK (2023) Examining the correlation between sexual and reproductive health stigmatization level and gender perception: a case of a university in Turkey-a descriptive cross-sectional study. Sao Paulo Med J 141(2):146–153
23. Hussein J, Ferguson L (2019) Eliminating stigma and discrimination in sexual and reproductive health care: a public health imperative. Sex Reprod Heal Matters 27(3):1–5

24. Inanc H, Meckstroth A, Keating B, Adamek K, Zaveri H, O'Neil S et al (2020) Factors influencing youth sexual activity: conceptual models for sexual risk avoidance and cessation. OPRE Res Br [Internet] 153:1–26. Available from: https://www.acf.hhs.gov/sites/default/files/documents/opre/factors-influencing-youth-sexual-activity-dec-2020_0.pdf
25. Ahinkorah BO (2020) Predictors of unmet need for contraception among adolescent girls and young women in selected high fertility countries in Subsaharan Africa: a multilevel mixed effects analysis. PLoS One [Internet] 15(8 Aug):1–15. Available from: https://doi.org/10.1371/journal.pone.0236352
26. Kumar A, Singh A (2013) Trends and determinants of unmet need for family planning in Bihar (India): evidence from national family health surveys. Adv Appl Sociol 03(02):157–163
27. Nyoman Sumiati L, Nyoman Wirawan D, Seri Ani L (2019) Determinants of unmet needs for family planning in Indonesia: secondary data analysis of the 2017 Indonesia demographic and health survey. Public Heal Prev Med Arch 2019 [Internet] 7(2):85–94. Available from: www.sdki.bkkbn.go.id
28. Bronfenbrenner, U. (1977). Toward an experimental ecology of human development. Am Psychol 32(7):513
29. Pasay-an E, Magwilang JOG, Pasay-an E, Magwilang JOG, Pangket PP (2020) Knowledge, attitudes, and practices of adolescents regarding sexuality and reproductive issues in the Cordillera administrative region of the Philippines. Makara J Heal Res 24(3)
30. American Academy of Pediatrics (AAP) (2021) Equitable access to sexual and reproductive health care for all youth
31. Morris JL, Rushwan H (2015) Adolescent sexual and reproductive health: the global challenges. Int J Gynecol Obstet 131:S40–S42
32. Roka ZG, Akech M, Wanzala P, Omolo J, Gitta S, Waiswa P (2013) Factors associated with obstetric fistulae occurrence among patients attending selected hospitals in Kenya, 2010: a case control study. BMC Pregnancy Childbirth 13:1–7
33. Zia Y, Mugo N, Ngure K, Odoyo J, Casmir E, Ayiera E et al (2021) Psychosocial experiences of adolescent girls and young women subsequent to an abortion in Sub-saharan Africa and globally: a systematic review. Front Reprod Heal 3(May)
34. Beacham BL, Deatrick JAMY, Elmore S (2008) 基因的改变NIH Public Access. Bone 23(1):1–7
35. Vohra R, Vohra A, Sharma S, Rathore M, Sharma B, Sharma M (2014) Determinants of the unmet need for family planning among women of Jaipur, Rajasthan. Int J Adv Med Heal Res 1(1):20
36. Rohmah N, Yusuf A, Hargono R, Laksono AD, Masruroh, Sujoso ADP et al (2021) Barrier to contraceptive use among childbearing age women in rural Indonesia. Malaysian Fam Phys 16(3):16–22
37. Nyblade L, Ndirangu JW, Speizer IS, Browne FA, Bonner CP, Minnis A et al (2022) Stigma in the health clinic and implications for PrEP access and use by adolescent girls and young women: conflicting perspectives in South Africa. BMC Public Health 22(1):1–11
38. Nmadu AG, Mohamed S, Usman NO (2020) Adolescents' utilization of reproductive health services in Kaduna, Nigeria: the role of stigma. Vulnerable Child Youth Stud [Internet] 15(3):246–56. Available from: https://doi.org/10.1080/17450128.2020.1800156
39. WHO, UNAIDS, UNFPA, UNICEF, UN WOMEN TWB (2018) Survive, thrive, transform current status and strategic priorities. Glob Strateg Women's, Child Adolesc Heal 15(Nov):68
40. Kemenkes RI (2020) Pedoman Pelayanan Antenatal Terpadu 2020 Ed.3 [Internet]. Kementerian Kesehatan Republik Indonesia. III:4–6. Available from: https://repository.kemkes.go.id/book/147
41. Kemenkes RI (2018) Hasil Riset Kesehatan Dasar Tahun 2018. Kementrian Kesehat RI. 53(9):1689–1699
42. Coulson J, Sharma V, Wen H (2023) Understanding the global dynamics of continuing unmet need for family planning and unintended pregnancy. China Popul Dev Stud [Internet] 7(1):1–14. Available from: https://www.ncbi.nlm.nih.gov/pmc/articles/PMC10075166/#CR45

43. Chandra-Mouli V, Plesons M, Hadley A, Maddaleno M, Oljira L, Tibebu S et al (2017) Lessons learned from national government led efforts to reduce adolescent pregnancy. Annu Meet Youth Advis Counc Heal pp 50–6
44. Chandra-Mouli V, Akwara E (2020) Improving access to and use of contraception by adolescents: what progress has been made, what lessons have been learnt, and what are the implications for action? Best Pract Res Clin Obstet Gynaecol [Internet] 66:107–118. Available from: https://doi.org/10.1016/j.bpobgyn.2020.04.003
45. Chandra-Mouli V, Parameshwar PS, Parry M, Lane C, Hainsworth G, Wong S et al (2017) A never-before opportunity to strengthen investment and action on adolescent contraception, and what we must do to make full use of it. Reprod Health 14(1):1–13
46. Huang KY, Kumar M, Cheng S, Urcuyo AE, Macharia P (2022) Applying technology to promote sexual and reproductive health and prevent gender based violence for adolescents in low and middle-income countries: digital health strategies synthesis from an umbrella review. BMC Health Serv Res [Internet] 22(1):1–27. Available from: https://doi.org/10.1186/s12913-022-08673-0
47. Coker AL, Davis KE, Arias I, Desai S, Sanderson M, Brandt HM et al (2021) Reprint of: physical and mental health effects of intimate partner violence for men and women. Am J Prev Med 61(6):777–786
48. Sinai I, Nyenwa J, Oguntunde O, Sinai I, Nyenwa J, Oguntunde O (2018) Programmatic implications of unmet need for contraception among men and young married women in northern Nigeria. Open Access J Contracept 9(81):81–90
49. Hailu S, Assefa N, Dingeta T, Abdurahman C, Adem M (2022) Unmet need for contraception among married adolescent girls and young women in Haramaya Health and demographic surveillance system, Eastern Ethiopia. Front Glob Women's Heal 3
50. Rizvi F, Williams J, Bowe S, Hoban E (2020) Factors influencing unmet need for contraception amongst adolescent girls and women in Cambodia. PeerJ 8:1–25
51. Buller AM, Schulte MC (2018) Aligning human rights and social norms for adolescent sexual and reproductive health and rights. Reprod Health Matters [Internet] 26(52):38–45. Available from: https://doi.org/10.1080/09688080.2018.1542914

Ni Komang Yuni Rahyani The author is a lecturer in the department of Midwifery who has worked for more than 30 years. Activities as a lecturer require unlimited creativity. The work that has been produced is in the form of articles in international journals, monographs, book chapters, modules, and textbooks, as well as intellectual property rights. This article is the second article in Springer. Hopefully, the works created can provide additional experiences for readers.

Health Equity for the Tribal Minority: Issues from Indochina

13

Rujittika Mungmunpantipantip and Viroj Wiwanitkit

Abstract

The difficulties and problems with health fairness for ethnic minorities in Indochina are examined in this chapter. In nations like Vietnam, Laos, Cambodia, and Myanmar, tribal minorities have significant disparities in healthcare access, leading to inferior health results in comparison to the majority population. The chapter examines the social, economic, and geographic factors—discrimination, remote areas, a lack of infrastructure, and cultural barriers—that lead to these disparities. The chapter also addresses the unique health problems faced by tribal minorities, including increased rates of infectious diseases, malnourishment, problems with the health of mothers and children, and insufficient access to essential medical care. It mentions how these health disparities affect the overall growth and well-being of indigenous communities. The chapter also focuses on a variety of initiatives and treatments that try to improve the health equality of indigenous minorities. Among them are projects aimed at enhancing the infrastructure of healthcare, expanding access to essential services, training healthcare professionals in cultural competency, and promoting community involvement and empowerment. To achieve long-term improvements in health equity, it also highlights how important cooperation between governments, non-governmental organizations, and tribal people is. In summary, this chapter clarifies the crucial health equity concerns that ethnic minorities in Indochina face, highlighting the significance of adopting all-encompassing and

R. Mungmunpantipantip (✉)
Private Academic Consultant, 111 Bangkok 122, Bangkok 103300, Thailand
e-mail: rujittika@gmail.com

V. Wiwanitkit
Chandigarh University, Ajitgarh, India

Joesph Ayobabalola University, Ikeji-Arakeji, Nigeria

Hainan Medical University, Haikou, China

B. W. K. Son (ed.), *The Landscape of Global Health Inequity*, Integrated Science 22,
https://doi.org/10.1007/978-3-031-60502-4_13

inclusive approaches to tackle these obstacles. The recognition and handling of the unique needs and rights of tribal minorities can pave the way for the development of a more just and inclusive healthcare system that ensures the health and dignity of every individual.

Keywords

Health • Equity • Tribal • Minority • Indochina

1 Introduction

Indochina is a specific region of the planet located in Asia and home to thousands of millions of people. The Southeast Asian region that consists of Vietnam, Cambodia, and Laos is known as "Indochina." The French colonies that existed in the area in the nineteenth and twentieth centuries are frequently referred to by this name. Numerous different ethnic groups call the Southeast Asian region of Indochina home. The Vietnamese, Lao, Thai, and Khmer (Cambodian) peoples are some of the major ethnic groups in the region. There are lesser ethnic groups including the Hmong, Cham, Mon, and several hill tribes in addition to these dominating populations. With their distinct cultures, languages, and customs, these ethnic groups contribute to Indochina's great diversity.

When health inequality is mentioned, it is important to talk about how different groups of people have varied health outcomes and access to healthcare. These discrepancies are usually caused by characteristics like income level, gender, race/ethnicity, education level, and geographic location. Due to these differences, some communities may be more vulnerable to disease, disability, and early mortality. Because it can have a substantial effect on people's individual and community well-being, health inequality is important. Access to high-quality healthcare services is frequently impeded for those who encounter health disparities, which can result in untreated or inadequately managed medical issues. Higher rates of morbidity and mortality, a decline in quality of life, and higher healthcare expenses can all arise from this. Due to its potential to exacerbate social and economic inequalities, health inequality also has wider societal ramifications.

A comprehensive approach is required to achieve health equity for the indigenous minority of Indochina. It means addressing the underlying social, economic, and cultural health determinants in addition to improving access to services and bolstering the healthcare infrastructure. To develop context-specific solutions that support inclusivity, cultural competency, and community empowerment, cooperation between governments, non-governmental organizations, and tribal groups is crucial. By highlighting the unique challenges and issues encountered by tribal minorities in Indochina and making the case for interventions and policies that prioritize the health and well-being of these underserved populations, this chapter seeks to further the continuing discussion on health equity.

2 Health Equity and Minority

Ensuring that every individual has an equal opportunity to achieve optimal health outcomes is the fundamental goal of health equality as a concept in public health practice. A multidisciplinary approach that considers social, economic, and environmental issues—all of which contribute to health inequities—is necessary to incorporate equity into public health practice. Over time, interdisciplinary contributions from fields including sociology, economics, and epidemiology have influenced the concept of health equity [1]. A multimodal approach is required to address health disparities in core public health functions [1]. It means bringing equitable concerns into responsibilities like policy development, assurance, health promotion, and health evaluation. Appropriate frameworks, including the Health in All Policies approach and the Social Determinants of Health model, offer guidance on how to integrate equity into public health practice [1]. A multimodal approach is required to address health disparities in core public health functions [1]. It means bringing equitable concerns into responsibilities like policy development, assurance, health promotion, and health evaluation. Appropriate frameworks, including the Health in All Policies approach and the Social Determinants of Health model, offer guidance on how to integrate equity into public health practice [1]. Measuring the advancement of health equity is challenging due to the complexity and multifaceted nature of health inequities [1]. It is often the case that existing health indicators fall short of fully capturing the extent of disparities, thereby calling for the development of new metrics and approaches to data collecting.

Health outcomes are significantly impacted by basic educational competency and skills, such as foundational knowledge, reasoning capacity, emotional self-regulation, and interactional capacities. People who receive an education are more equipped to make health-related decisions, adopt healthy behaviors, and deal with healthcare systems. Education also fosters problem-solving skills, critical thinking, and emotional stability—all of which support resilience and general health. Significantly, health outcomes are influenced by education, which is thought to be a fundamental social predictor of health. Policies and programs that decrease the attainment disparities in education between various populations are necessary to advance health equity. The main objectives of these interventions ought to be to guarantee equitable access to high-quality education for underprivileged groups, guarantee the availability of resources and support, and remove structural obstacles that hinder academic achievement [2]. Public health officials, healthcare professionals, educators, and departments of health and education must work together to execute evidence-based educational initiatives and policies that have measurable positive effects on public health [2].

It may make it more difficult to receive necessities like healthcare, education, and sanitary facilities. Communities in these locations may experience higher rates of poverty, hunger, and a general decline in quality of life as a result [1, 2]. Furthermore, inadequate infrastructure might impede the efficient operation of businesses and impede the progress of economic development in these areas [1, 2]. In general, removing these obstacles is essential to advancing equitable development and enhancing everyone's quality of life, wherever they may be. Minorities confront significant health disparities, which should be highlighted (Table 1). Globally, minorities are confronted with several significant health inequalities that affect their overall health outcomes, the prevalence of certain health disorders, and access to healthcare. One major problem is the lack of access to healthcare, especially in rural areas where infrastructural and physical obstacles make it difficult for people to get the treatment they need. Minority communities may face challenges in accessing healthcare services due to language problems and cultural insensitivity on the part of healthcare practitioners [1, 2]. Furthermore, a large number of minorities are uninsured or have little financial resources, which keeps them from receiving necessary medical care [1, 2].

Apart from facing obstacles in obtaining healthcare, minority groups also face a greater incidence of particular medical disorders in comparison to the overall population. Minorities are more likely to suffer from chronic illnesses like diabetes and cardiovascular disease as well as infectious diseases like HIV/AIDS and tuberculosis. Minority groups also experience higher rates of mental health disorders, malnutrition, and micronutrient deficiencies, which adds to the overall health inequities in these populations. Minority health inequalities are largely influenced by socioeconomic variables as well. A number of factors, including poverty, a lack of education, discrimination against, and social isolation from minority groups, can lead to poor health outcomes. Minority populations are disproportionately affected by environmental hazards and occupational dangers, which exacerbate health disparities and restrict access to essential healthcare services. Achieving health equity for all people, regardless of race or origin, requires addressing these complicated concerns.

3 The Health and Development Concerns of Tribal Communities

The health and development concerns of tribal communities are multi-faceted and require comprehensive attention. Some key concerns include the following issues. First, tribal tribes frequently live in isolated, remote locations far from medical facilities. Due to this geographic barrier, people have less access to basic healthcare services like treatment for common illnesses and chronic ailments, maternity and child health services, and preventative care. Second, health disparities exist. Compared to the mainstream population, tribal societies have greater rates of infectious diseases, poverty, and malnutrition. Social determinants of health, such as poor access to sanitary conditions, clean water, education, and work opportunities, are frequently the cause of these disparities. Third, cultural and linguistic barriers are common. Due to their unique cultural customs and languages, tribal communities may face difficulties receiving healthcare. Effective healthcare delivery can be hampered by poor communication, a shortage of culturally competent healthcare professionals, and a lack of knowledge about the treatments offered. Fourth, tribal communities often rely on traditional healing practices and herbal medicines.

Table 1 Key health disparities faced by minorities

Problems	Management
Limited access to healthcare	Access to healthcare is a critical issue for many people around the world, with geographical constraints and insufficient infrastructure in remote locations making it difficult for individuals to access the care they require. Furthermore, language obstacles and cultural insensitivity from healthcare personnel can discourage people from obtaining medical attention. Lack of health insurance or budgetary restraints can provide substantial problems, preventing many people from receiving necessary healthcare services. These impediments cause inequities in healthcare access, resulting in disparities in health outcomes and worsening the health gap between various groups. Efforts to resolve these difficulties are critical to ensure that all people have equal access to healthcare, regardless of geography, language, culture, or financial situation
High prevalence of some specific health problems	Various factors influence the prevalence of specific health conditions in populations with socioeconomic disparities, including infectious diseases like HIV/AIDS and tuberculosis, chronic diseases like diabetes and cardiovascular disease, mental health issues, malnutrition, and micronutrient deficiencies. These differences are worsened by socioeconomic variables such as poverty and low education, discrimination and social isolation, as well as exposure to environmental and occupational hazards. Populations confronting these issues frequently have limited access to adequate healthcare, good food options, and safe living conditions, which increases vulnerability to a variety of health conditions and lowers overall well-being
Socioeconomic factors contributing to disparities	Poverty and restricted education, discrimination and social exclusion, as well as environmental and occupational risks, all contribute to societal inequities. Individuals living in poverty frequently struggle to get resources and opportunities that can enhance their quality of life, such as healthcare, education, and housing. Limited education can also prolong poverty cycles, as people with higher levels of education are more likely to have secure jobs and higher salaries. Discrimination and social exclusion can further marginalize specific groups of people in society, prohibiting them from fully engaging in economic, political, and social activities. Furthermore, environmental hazards and occupational risks might disproportionately affect low-income communities and workers, resulting in health inequities and a lower quality of life. To ensure equitable access to healthcare and resources, it is necessary to address these gaps through focused interventions that address both the health conditions and the underlying social determinants

While these practices hold cultural significance, there is a need to ensure the safety and effectiveness of such treatments and to integrate traditional healing practices into the broader healthcare system. Last, land rights and displacement as a result of development projects, conflicts, or environmental degradation are issues that many indigenous tribes must deal with. These problems have the potential to worsen health disparities and upend traditional livelihoods.

A multi-sectoral approach combining cooperation between government agencies, civil society organizations, and tribal community leaders is necessary to address the health and development concerns of tribal populations. Enhancing access to healthcare through telemedicine services, community health workers, and mobile clinics should be the main goal of strategies [4]. It is important to support healthcare services that are culturally aware and include conventional therapeutic methods. To further support sustainable development and fair health outcomes for tribal people, efforts must be made to address the underlying social determinants of health, such as poverty, education, and land rights [4].

4 Ethnic Minority in Indochina: A Large Group that is Forgotten for the Good Public Health System

When it comes to having access to a robust public health system, ethnic minorities in Indochina are often disregarded, despite their size. At present, the minorities have several basic problems due to (a) primitive lifestyle, rooted ritual culture and belief, (b) remote areas, access limitations, difficulty in approach, (c) environment that might promote the growth of pathogens, and (d) poor sanitation system, no necessary health facility. Due to this neglect, these people have unequal access to healthcare and poorer outcomes, which has created a vicious cycle of ill health with few opportunities for improvement. To address this issue, it is necessary to acknowledge the distinct healthcare needs and cultural contexts of ethnic minority groups. Additionally, targeted interventions and policies that prioritize health equity and guarantee that no one is left behind in the pursuit of robust public health system benefits must be implemented.

Indigenous tribes in Indochina, such as the Hmong in Laos, confront a wide range of health issues as a result of displacement and the loss of traditional customs. The forcible transfer of these people frequently disrupts their traditional way of life, including access to traditional healing techniques and knowledge. This can lead to poorer health outcomes because individuals may lack access to tools and activities that are fundamental to their cultural health beliefs and practices. Furthermore, indigenous people frequently face barriers to accessing culturally competent healthcare services since many healthcare providers do not understand or respect their cultural beliefs and traditions. This lack of awareness can result in misdiagnosis, insufficient treatment, and overall inferior health outcomes for people of these minority groups. Furthermore, linguistic obstacles and systemic prejudice can impede their access to adequate healthcare, worsening their health issues.

The Hmong, Karen, Yao, and Akha hill tribes, as well as the Moken and Chao Lay sea kayak groups, are a few examples of ethnic tribal minority groups in Indochina. Many ethnic minority groups that inhabit the mountainous areas of the Indochinese Peninsula countries, including Vietnam, Laos, and Thailand, are referred to as "hill tribes" in Indochina. These groups are frequently referred to as hill tribes because of their historical settlements in the highlands. One of the biggest ethnic minority groups in Vietnam and Laos, the Hmong also make up a sizeable portion of the populations of China and Thailand. Several million Hmong people are thought to reside in these nations, while precise statistics are difficult to come by. The basic problems for the sea kay groups include limited access to healthcare services, lack of clean water and sanitation, malnutrition and food insecurity, infectious diseases such as malaria and tuberculosis, maternal and child health issues, substance abuse, and environmental health risks. The sea kayak groups, also called sea gypsies, are a minority ethnic group that primarily inhabits the coastal regions of Thailand and Myanmar (Burma). The Moken people are found nearby, however, they are not officially in Indochina, the term used to describe Vietnam, Laos, and Cambodia. The Moken lead a unique way of life that is oriented around the sea. They are a group of seafarers renowned for their exceptional prowess in diving, fishing, and water navigation. Because they depend on the ocean for their food and survival, they have a keen understanding of it and its resources. The basic problems for the sea kay groups include limited access to healthcare services, lack of clean water and sanitation, malnutrition and food insecurity, infectious diseases such as malaria, maternal and child health issues, and environmental health risks:

These communities frequently live in distant, rural locations and have particular difficulties accessing essential amenities like healthcare and education. Their well-being is contingent upon the acknowledgment and incorporation of their needs into public health policies and interventions, given the distinctiveness of their cultural traditions, languages, and livelihoods. Due to the poor local statistical surveillance system in remote areas of Indochina, especially for the areas with local conflict, the exact number of minorities in this area is currently not definitively known. However, in some specific areas with the systematic collection of statistical data, it seems that there are up to millions of monitories in Indochina. For example, data from the Social Development Unit in Thailand [5], the statistics on the population residing in highland communities in Thailand, categorized by ethnic groups in 20 provinces, show that there is a total of 923,257 people, 164,637 households, and 186,413 families belonging to 10 different hill tribes. Among these hill tribes, the most populous group is the Karen tribe with 438,131 people. Classified by groups of hill tribes, Karen comprises the highest percentage (about 37%) followed by Mhong about 8%). This is a significant number considering that the overall population in Thailand is 65 million.

Access to healthcare for Indochina's tribal minorities is hampered by a lack of infrastructure, geographical isolation, and cultural and linguistic disparities. These settlements are frequently located in distant alpine regions or border regions, far from healthcare facilities and services. Due to a lack of transportation and communication infrastructure, ethnic minorities find it difficult to reach healthcare facilities, resulting in limited access to basic healthcare services. Cultural and language limitations also impede tribal populations' access to and utilization of healthcare. Traditional beliefs, customs, and linguistic variations can cause communication gaps between healthcare practitioners and tribal groups, resulting in misconceptions and poor care quality. Discrimination and stigma compound these difficulties, with indigenous minorities frequently experiencing marginalization and unequal treatment in the healthcare system.

It's noteworthy that there is still a conflict zone in Indochina, and this could pose a serious issue for the minority who reside there. (such as areas with internal fighting between government and minorities). When it comes to getting healthcare services, ethnic minorities in Indochina's conflict zones—like the Hmong, Karen, and Rohingya communities—face considerable obstacles. Access to vital healthcare facilities and supplies is frequently restricted or nonexistent in these areas because of the continuous hostilities and instability. In addition, obstacles can prevent them from getting the medical attention they need for their wounds because hospitals could be destroyed or hard to reach as a result of fighting. Ethnic minorities may also be more susceptible to infectious diseases in conflict-affected areas due to poor sanitation and hygiene. In addition to causing overcrowding, a lack of clean water sources, and a shortage of medical supplies, population displacement in these areas can also spread diseases like respiratory infections, malaria, and cholera. Ethnic minorities in Indochina's conflict zones have a difficult time getting access to healthcare, which can be harmful to their general health and wellness. To meet these populations' specific healthcare needs and advance their general well-being, efforts must be made to increase their access to healthcare [6].

Interestingly, the issue is highly complex in the isolated border region. The Golden Triangle is a region that is well-known worldwide [7]. The region in Indochina where the borders of Thailand, Laos, and Myanmar converge is known as the "Golden Triangle." It is well recognized for its past connections to the opium trade and for being a hub for drug manufacturing and trafficking. The area was once a significant supplier of illegal drugs, and its opium crop is where the moniker "Golden Triangle" originates. These days, the name covers a wider area that includes portions of all three countries, which are renowned for their breathtaking scenery, diverse ethnic populations, and rich cultural histories. Due to its natural beauty, cultural significance, and historical relevance, the Golden Triangle has become a more well-liked tourist destination [7]. There is a tribe living in this rough, primarily forested area, which is home to between 300,000 and 500,000 people living in roughly 3000 communities [7, 8]. The people who live seminomadic lifestyles carry weapons and personal items with them at all times. They have been farming opium poppies since the turn of the century, and this harvest provides for all of the families' dietary needs [7, 8]. In an area where backpack transportation is the primary mode of transportation, opium provides a good return for a low weight. In addition to being the region's source of drug addiction—which is perhaps more accurately described as the local remedy for the discomforts of illness—opium is utilized to supply the world market [7].

5 Important Public Health Problems for Tribal Minority Groups in Indochina

Tribal minority groups in Indochina face a number of grave public health problems that require consideration and treatment. These are but a handful of instances. Due to their often remote and rural locations, tribal villages sometimes have limited access to medical facilities. Inequities in healthcare services, including vaccinations, treatment for common illnesses and chronic conditions, and preventative care, are brought about by this inaccessibility. Among tribal cultures, infectious diseases like malaria, tuberculosis, and watery illnesses might be more common. The occurrence and consequences of these illnesses are influenced by inadequate hygiene, a scarcity of potable water, and an inadequate healthcare system. Poverty, especially among women and children, a lack of agricultural resources, and restricted access to nutrient-dense foods all aggravate inadequate diets and nutritional deficiencies. Tribal groups often have higher rates of maternal and child mortality than the general population. Cultural practices and beliefs, as well as limited access to prenatal care, skilled delivery attendants, and postnatal care, can all worsen the health outcomes for mothers and their children.

Tribal groups may face unique mental health issues as a result of social and cultural conditions, displacement, and exposure to trauma. The stigma associated with mental illness and the lack of access to mental healthcare services in many areas contributes to the underdiagnosis and undertreatment of mental health problems. A comprehensive strategy is needed to address these public health issues, including addressing food insecurity, boosting access to critical healthcare services, enhancing healthcare infrastructure, and offering community-based and culturally sensitive interventions. Promoting the welfare of tribal minority communities in Indochina and effectively resolving these public health issues require cooperation between governments, non-governmental organizations, and leaders of the tribal communities. Here, a few significant illnesses and public health issues that still affect Indochina's tribal minority communities will be shown.

5.1 Malaria

The Plasmodium parasite is the cause of malaria, an infectious disease that is carried by mosquitos. Humans contract it through the bites of female Anopheles mosquitoes carrying the infection. Malaria is a serious global health concern, especially in tropical and subtropical countries of Africa, Asia, and Latin America. Fever, headaches, chills, and flu-like symptoms are some of the signs of malaria. An estimated 249 million cases of malaria were reported worldwide in 2022, with

608,000 malaria deaths reported in at least 85 countries. About 80% of all malaria deaths in the Region occurred in children under the age of five [9].

If left untreated, malaria can result in life-threatening illness and even death, particularly in young children, expectant mothers, and those with weakened immune systems. Human malaria is caused by various kinds of Plasmodium parasites, the most deadly of which is Plasmodium falciparum. Laboratory techniques that identify the parasite's presence in blood samples can identify malaria. A range of tactics is used to prevent and control malaria, including the use of insecticide-treated bed nets, indoor residual pesticide spraying, antimalarial medication for both treatment and prevention, and attempts to eradicate mosquito breeding grounds. Researchers are developing new antimalarial drugs, improved diagnostic tools, and vaccinations to treat the condition. To lessen the effects of malaria and enhance public health in impacted areas, eradication initiatives are essential. This includes vector control strategies, community education and awareness campaigns, and early case detection and treatment.

The frequency of malaria infection in Indochina's minority communities is influenced by multiple factors. Incidence rates of malaria in Indochina varied from 125.58 to 60,577 per 10,000 people [10]. The highest incidence occurs in the remote border area between countries in Indochina and the problem of drug resistance is common in this area [10]. Minority populations in Indochina are often found in remote, underprivileged areas with little access to medical care. Socioeconomic challenges such as destitution, substandard housing, and limited access to sanitary facilities and safe drinking water could befall them [11, 12]. These variables all contribute to their increased susceptibility to malaria infection. In Indochina, minority communities frequently lack access to healthcare, including diagnostic tests, antimalarial drugs, and mosquito control methods [11, 12]. This lack of access impedes early diagnosis and timely treatment, resulting in higher morbidity and mortality rates. Due to factors such as deforestation, closeness to stagnant water sources, and poor housing conditions, minority people may dwell in locations with a high frequency of malaria-carrying mosquitoes [11, 12]. Additionally, the inappropriate distribution of public health personnel and facilities still exists. This inequity results in the problem of malaria control in poor areas of Southeast Asia [13, 14]. As a result, the emerging multidrug-resistant malaria strain occurs and becomes the new public health threat.

In these societies, several environmental factors raise the risk of malaria transmission. Cultural beliefs and practices: Minority communities' cultural practices and beliefs can influence their understanding and perception of malaria prevention and treatment. Traditional attitudes and practices may conflict with modern preventative measures, resulting in limited adoption of interventions such as insecticide-treated bed nets or antimalarial drug adherence [11, 12]. Due to conflict, several minority communities in Indochina have been uprooted, resulting in a breakdown in healthcare infrastructure and impaired access to key services. This displacement may make them more vulnerable to malaria infection. Addressing the malaria infection problem among Indochina's minority communities necessitates a focused and culturally appropriate approach [13]. This includes boosting access to healthcare services, as well as promoting malaria education and awareness. This involves improving access to healthcare, educating people about malaria prevention and control techniques, and tackling the underlying social determinants of health such as poverty and displacement [11, 12]. In these societies, several environmental factors raise the risk of malaria transmission. Cultural beliefs and practices: Minority communities' cultural practices and beliefs can influence their understanding and perception of malaria prevention and treatment. Traditional attitudes and practices may conflict with modern preventative measures, resulting in limited adoption of interventions such as insecticide-treated bed nets or antimalarial drug adherence [11, 12]. Due to conflict, several minority communities in Indochina have been uprooted, resulting in a breakdown in healthcare infrastructure and impaired access to key services. This displacement may make them more vulnerable to malaria infection. Addressing the malaria infection problem among Indochina's minority communities necessitates a focused and culturally appropriate approach [13]. This includes boosting access to healthcare services, as well as promoting malaria education and awareness. This involves improving access to healthcare, educating people about malaria prevention and control techniques, and tackling the underlying social determinants of health such as poverty and displacement [11, 12]. To effectively combat malaria in these vulnerable groups, collaborative measures involving governments, healthcare providers, community leaders, and non-governmental organizations are required.

5.2 Tuberculosis

Mycobacterium tuberculosis is the bacterium that causes tuberculosis (TB). TB is still a widespread illness that affects people worldwide. More than 99 percent of the world's population lives in 192 countries and regions, which supplied statistics on tuberculosis cases in 2023 [14]. Although the lungs are the main organs affected, the kidneys, spine, and brain may also be affected. The bacteria are breathed by others and dispersed through the air when an infected person coughs, and sneezes. For several reasons, tuberculosis (TB) is a serious public health concern in Indochina. The region of Indochina, comprising Vietnam, Laos, Myanmar, Thailand, and Cambodia, has a high rate of tuberculosis prevalence. It is estimated that 322 cases of bacteriologically confirmed tuberculosis (TB) per 100,000 people live in Indochina [15]. Also, the problem of drug resistance is common in this area [15]. The area is known as a tuberculosis (TB) hotspot due to a number of variables, including high rates of risk factors like HIV/AIDS and malnutrition, poverty, overcrowding, and limited access to healthcare. The spread of drug-resistant tuberculosis (TB) in Indochina, especially extensively drug-resistant (XDR-TB) and multidrug-resistant (MDR-TB), is a serious issue [16–19]. Drug-resistant strains of tuberculosis are more challenging to treat and require longer treatment courses with more costly and harmful drugs. HIV/AIDS is comparatively common in Indochina, where TB-HIV co-infection is a serious public health issue. HIV weakens the immune system, increasing a person's susceptibility to contracting tuberculosis and increasing the chance that the disease may advance and cause death [18]. In Indochina, many individuals have obstacles in accessing healthcare, especially in remote and marginalized areas. Effective management of tuberculosis is hampered by inadequate treatment, delayed diagnosis, and limited diagnostic facilities [17–19]. Additionally, because stigma can cause people to fear social rejection or isolation, it might impede efforts to diagnose and treat tuberculosis cases. Strengthening healthcare systems, expanding access to high-quality treatments and diagnostic tools, addressing social determinants of health, and promoting public awareness and education are all part of Indochina's efforts to battle tuberculosis (TB) [17–19]. Collaboration between governments, healthcare providers, non-governmental organizations, and foreign partners is essential to effectively combat tuberculosis and lower its incidence in the region [17–19].

Currently, the unique issues of TB care among ethnic minority people in Indochina are fraught with numerous significant challenges. In Indochina, tribal minority communities are often situated in secluded and remote areas, making access to medical facilities difficult. Their capacity to seek prompt diagnosis and treatment for tuberculosis is hampered by this lack of access. Long travel times to medical institutions, a lack of transportation, and a poor infrastructure for the medical field all contribute to this issue [11, 12]. It may be challenging to give tribal minority groups the proper TB care due to linguistic and cultural barriers. Problems in communication between patients and medical staff can make it difficult for patients to receive the proper diagnosis, adhere to their treatment plans, and understand efforts to prevent and control tuberculosis [11, 12]. Culturally competent approaches—like employing community health workers and interpreters—are essential to overcoming these challenges. TB is associated with stigma and discrimination because of cultural practices and beliefs; this can be particularly evident in tribal minority communities [11, 12]. The Golden Triangle, a rural area of Indochina that is home to several hill tribal cultures, has a high rate of tuberculosis despite this, according to a recent report [18]. This is an extremely challenging issue. Local conflicts and wars persist, and it is undeniable that disease management in a fighting zone is unachievable [18]. Culturally competent approaches—like employing community health workers and interpreters—are essential to overcoming these challenges. TB is associated with stigma and discrimination because of cultural practices and beliefs; this can be particularly evident in tribal minority communities [11, 12]. The Golden Triangle, a rural area of Indochina that is home to several hill tribal cultures, has a high rate of tuberculosis despite this, according to a recent report [18]. This is an extremely challenging issue. Local conflicts and wars persist, and it is undeniable that disease management in a fighting zone is unachievable [18]. Lessons from the conditions of the past Cambodian war provide a good explanation for the current problem of minority conflict in the Golden Triangle [20, 21]. To ensure that ethnic minorities receive prompt and adequate TB care, it is imperative to combat stigma through community education and awareness campaigns [11, 12]. Tribal minority societies probably have relatively little knowledge of tuberculosis symptoms, transmission, and treatment options. Low educational attainment and low health literacy may be the cause of treatment delays and poor adherence. Health education campaigns led by well-respected community members and leaders can contribute to greater public knowledge and awareness of TB. The cultural context should inform these programs.

5.3 HIV Infection

HIV infection is a virus that attacks the immune system, most especially the CD4 cells (T cells), which are vital for fighting diseases and infections. The most extreme form of HIV infection, known as AIDS (Acquired Immunodeficiency Syndrome), can develop if the virus is left untreated. Official data from the World Health Organization indicates that up to 45.7 million people worldwide are expected to be HIV-positive in 2023 [22]. The highest prevalent rates are currently reported from developing countries in Africa and Asia [23]. Despite numerous obstacles, HIV infection is still a significant public health concern in Indochina. Despite numerous obstacles, HIV infection is still a significant public health concern in Indochina. In Indochina, HIV infection is comparatively widespread, especially in certain demographics that are more vulnerable, such as transgender individuals, injectable drug users, men who have sex with men, and sex workers. In Indochina, there were almost 0.09 new HIV cases for every 1,000 people who were at risk in 2021 [24]. The sexually active population, aged 15–49 for both sexes, is the most affected [24].

The transmission of HIV is facilitated by high-risk behaviors, stigma, and limited access to prevention interventions [19, 25]. HIV and tuberculosis (TB) are highly prevalent in Indochina. Diagnostic, treatment, and management challenges are unique when TB and HIV co-infect. One of the leading causes of death in the area for individuals living with HIV is tuberculosis. In Indochina, access is still a barrier despite advancements in the expansion of HIV prevention and treatment facilities [19, 25]. HIV testing, antiretroviral therapy (ART), and prophylactic supplies (such as condoms and clean needles) are insufficient. It is especially difficult for underprivileged groups, such as ethnic minorities and rural areas, to access these services. HIV-related prejudice and stigma persist in Indochina, which causes fear, social exclusion, and a reluctance to get tested and receive treatment. Effective HIV preventive programs, treatment compliance, and care engagement can all be hampered by stigma [19, 25]. There is a lack of infrastructure and healthcare services in several sections of Indochina, especially in remote and rural areas. The quality and accessibility of HIV prevention, testing, and treatment services may be impacted by this [19, 25].

For a number of reasons, the HIV epidemic's current relevance to ethnic minority populations in Indochina is remarkable. Due to their frequent social and economic marginalization, communities of tribal minorities are more susceptible to HIV infection. To evaluate the situation and identify trends and create a workable model for providing care during the early stages of HIV/AIDS infection to prevent subsequent TB infection, Apidechkul recommended that a TB and HIV/AIDS surveillance system be built for Hill-tribe communities [26]. Adolescents from Indochina's tribe are susceptible to STIs such as hepatitis B virus (HBV) and hepatitis C virus (HCV) infections because of their risk factors and sexual behaviors [27]. To lower the frequency of these diseases in the future, targeted and successful behavioral interventions are required, as these risk behaviors can vary between genders and tribes [27]. Tribal minority cultures may experience significant stigma and prejudice with HIV because of cultural beliefs, societal customs, and a lack of knowledge about the infection. Stigma can make people less likely to disclose their status, receive treatment, or get tested for HIV, which raises the risk of HIV spreading throughout the community [19, 25–27]. Tribal minority populations in Indochina often lack access to adequate healthcare resources, including HIV testing, counseling, and antiretroviral therapy (ART). Tribal minority HIV-positive individuals may thus have poorer health outcomes from a delayed diagnosis and insufficient treatment. Tribal minority communities in Indochina may face a variety of challenges, such as marginalization, limited access to education, and poverty. These challenges may increase the impact of the HIV epidemic and increase the difficulty for members of ethnic minorities to prevent, treat, and live with HIV infection [25–27].

Inadequate access to economic opportunities, healthcare, and educational opportunities can result in risky behaviors like drug use, sex work, and unprotected sex, which raises the risk of HIV transmission. Remote and vulnerable tribal minority communities sometimes lack access to HIV prevention initiatives. This includes providing condoms, clean needles for drug use, and information on safe sex practices. The spread of HIV among these communities may be facilitated by a dearth of targeted HIV prevention initiatives that are tailored to the unique needs and cultural context of tribal minorities [19, 21].

5.4 Narcotic Drug Use

The Golden Triangle area of Indochina, which borders Thailand, Laos, and Myanmar, is a major hub for drug manufacture and trafficking. The area is well-known for growing opium poppies, which are used to make heroin and other illegal substances [7, 8]. A sizable portion of the population in Indochina uses drugs, both injecting and not injecting. Drug abuse, including the use of heroin, methamphetamine, and synthetic opioids, is common and has a number of negative social and health effects. Because injectable drugs use shared needles, injection drug users are more likely to contract HIV [7, 8]. This adds to the region's HIV and other bloodborne infection epidemic. Drug usage is also associated with a number of health hazards, such as overdosing, mental health issues, and a higher susceptibility to other infectious diseases. Access to harm reduction strategies, including needle and syringe programs, opioid substitution therapy, and HIV testing and counseling, is severely limited despite efforts to expand preventive and treatment services. The region's ongoing drug-related harms are exacerbated by the limited provision of these services [7, 8, 28]. Drug users frequently experience stigma and discrimination, which can make it difficult for them to get social assistance and medical care. This makes dealing with drug use and its effects much more difficult [8]. The problem's transnational dimension lies in the strong connections between the drug trade in Indochina and both regional and global networks. This presents difficulties for law enforcement and necessitates international cooperation to successfully combat drug trafficking [28].

The usage of narcotic drugs by tribal minority communities in Indochina is a major concern. These populations frequently experience particular obstacles and vulnerabilities that contribute to drug use and the problems that accompany it [7, 8, 28]. Here are some crucial points concerning Indochina's indigenous minority and narcotics problem. In Indochina, tribal minority communities are frequently marginalized, discriminated against, and have restricted access to essential amenities such as healthcare, education, and employment prospects. Individuals may turn to drugs as a coping mechanism or to escape the hardships they confront as a result of these social and economic variables [7, 8, 28]. Tribal minority communities may be unaware of the risks and effects of drug usage because of their isolated settings and restricted access to information.

Prevention programs targeted at these groups may be more challenging to implement or culturally inappropriate, which causes a lack of knowledge about the risks associated with drugs. Certain populations of ethnic minorities reside in or close to regions known for drug production and trafficking, including the Golden Triangle [7, 8, 28]. Narcotics are more easily accessible and available due to their proximity to these locations, which facilitates the spread of drug use throughout these areas. It is common for tribal minority populations to lack access to appropriate medical resources and infrastructure, including drug treatment and rehabilitation programs [7–9, 28]. In addition to opium, salicylate analgesics were used by 75% of them [29]. Their addiction was most frequently caused by illness, particularly stomach pain [29]. makes it challenging for those who are addicted to drugs to get the help and therapy they need. Many indigenous minority communities have a long history of using drugs for cultural or spiritual purposes. These cultural practices can raise the likelihood of drug use and addiction when combined with extrinsic factors like poverty and limited opportunities [7, 8, 28].

6 Problems of Health Equity Among Tribal Groups in Indochina

In Indochina, the problem of health disparities between ethnic groups continues to be a major cause for concern. Health disparities arise from the fact that tribal communities in the region sometimes face many obstacles when trying to receive healthcare services. The following are some key points about health inequalities amongst ethnic groups in Indochina.

6.1 Limited Access to Healthcare

Access to healthcare facilities, including primary care clinics, hospitals, and specialist treatments, is sometimes restricted for tribal populations residing in remote and neglected areas. A lack of transit options, inadequate infrastructure, and remote location are all significant factors. Limited access to healthcare services impairs their ability to receive timely and adequate medical care. The hill tribe communities living in isolated places are a prime example of this issue [30]. Enhancing the availability of public health care for tribal members requires a deeper understanding of their cultural values, interests, and concerns. At the individual, community, and health system levels, comprehensive transformation is needed to provide culturally appropriate health education and adaptable service delivery options [30]. The resources and support available to volunteer health professionals in this remote area are limited [31]. Therefore, local governments will still need to help to address the current issues.

6.2 Socioeconomic Disparities

Tribal groups in Indochina often live in poverty with few economic opportunities. Their inability to afford healthcare, wholesome food, and other essential resources for optimal health contributes to health inequities. This problem is comparable to that which Maori and Pacific Islander peoples in New Zealand and Aboriginal and Torres Strait Islander peoples in Australia are facing [32].

6.3 Cultural and Language Barriers

Barriers to healthcare access may arise from cultural and linguistic differences between healthcare practitioners and indigenous populations. Native Americans sometimes struggle to adequately communicate their health concerns due to language barriers, which can lead to incorrect diagnoses or subpar treatment. Additionally, cultural differences may lead to a lack of cultural competence on the part of healthcare professionals, which may prevent them from understanding tribal people and providing them with proper care. Providing adequate public health care to tribal communities is hampered by a number of factors, including distance, socioeconomic deprivation, lack of cultural sensitivity, communication difficulties, tradition, beliefs, and differences in cultural practices [33], according to a Thai report. To provide healthcare, education, and information that is appropriate for the hill tribal people's culture and religion, healthcare services and locations must be modified [33].

6.4 Discrimination and Stigmatization

In Indochina, stigma and discrimination against tribal communities are commonplace, and this can result in unfavorable attitudes and biases among medical professionals. This can exacerbate health inequities by making Indigenous people less trusting and less likely to seek medical attention. Limited access to health education and awareness programs: Tribal groups may lack these resources, which might result in a lack of knowledge about healthy lifestyle choices, illness early detection, and prevention actions. Tribal groups may thus experience poorer health outcomes as a result of delayed diagnosis and treatment. For instance, the study discovered that common coping strategies among tribe members living with HIV are self-distancing, self-blame, and self-stigmatization [34]. Tribal tribes in Indochina face health disparities that must be addressed with a comprehensive strategy that considers their unique cultural, social, and economic situations. Enhancing healthcare services in remote and underserved areas for indigenous tribes and expanding healthcare infrastructure are two of the most crucial tactics.

7 Effective Tools to Manage Health Equity for Tribal Minority in Indochina

To control health disparities for ethnic minorities in Indochina, a number of useful instruments are available (Table 2).

7.1 Health Equity Assessment

Assessments of health equity can help pinpoint the particular health disparities and injustices that indigenous minority groups experience. This comprises collecting and examining information on social determinants of health, healthcare access, and other relevant aspects. The assessment can help to direct the creation of specific initiatives and policies to address observed disparities [28]. This may increase the likelihood of disease and unfavorable health outcomes, particularly for populations who may already be disproportionately vulnerable as a result of lifelong exposure to the environment and social factors [28].

7.2 Community Engagement and Empowerment

Reducing health disparities may be achieved by involving tribal minority communities in decision-making and giving them the authority to take an active role in their treatment. This can include assisting community-led projects and capacity-building efforts, as well as including the community in the planning,

Table 2 Using conceptual management to address major public health issues among Indochina's ethnic minority communities who have health inequity problems

Problems	Management
Malaria	Here are some key components of the Health Equity Framework that can be applied 1. Targeted Interventions 2. Access to Healthcare 3. Health Education and Community Engagement 4. Addressing Social Determinants of Health 5. Intersectoral Collaboration 6. Data Collection and Monitoring
Tuberculosis	Here are some key components of the Health Equity Framework that can be applied 1. Strengthening Health Systems 2. Community-Based Approaches 3. Early Case Detection: Enhance 4. Treatment Adherence Determinants 5. Collaboration and Partnerships 6. Monitoring and Evaluation Support 7. Addressing Social
HIV infection	Here are some key components of the Health Equity Framework that can be applied 1. HIV Education and Awareness 2. Access to HIV Testing and Counseling 3. Prevention of Mother-to-Child Transmission (PMTCT) 4. Condom Promotion and Distribution 5. Harm Reduction Interventions 6. Addressing Stigma and Discrimination 7. Strengthening Health System 8. Community Engagement and Empowerment:
Narcotic drug use	Here are some key components of the Health Equity Framework that can be applied 1. Prevention Programs 2. Early Intervention and Screening 3. Accessible Treatment Services 4. Harm Reduction Strategies 5. Community Engagement and Empowerment 6. Addressing Social Determinants 7. Collaboration and Partnerships 8. Monitoring and Evaluation

implementation, and assessment of healthcare programs [29]. Regarding community engagement and empowerment, the previous report from US can be a good model that can be applied to any setting including Indochina [29].

7.3 Culturally Competent Care

Addressing health disparities requires providing tribal minority groups with appropriate healthcare services that are culturally responsive. This involves educating medical professionals about the traditional beliefs, practices, and preferences of native communities. It also means accepting traditional treatment methods when appropriate and making sure that services are linguistically accessible. Research has shown that culturally appropriate therapy can be beneficial for ethnic minority groups with health concerns [30].

7.4 Health Education and Promotion

Tribal minorities can benefit from increased awareness of important health issues, preventative measures, and accessible healthcare facilities through the implementation of targeted health education and promotion initiatives. A Thai report states that to improve the health and well-being of hill tribe members and hence lessen health concerns, appropriate public health interventions are desperately needed [31].

7.5 Strengthening Health Systems

Managing health disparities requires bolstering healthcare services in regions inhabited by indigenous minorities. Public healthcare systems can be strengthened by the application of partnership-based and systematic strategies [32].

7.6 Intersectoral Collaboration

The creation and execution of policies that particularly address health disparities among tribal minorities are necessary for long-term management. For instance, the prior study demonstrated that the core of intersectoral action to prevent vector-borne illnesses is cooperation amongst several stakeholders to organize, carry out, and assess initiatives at various levels of intervention [33].

7.7 Policy and Advocacy

The creation and execution of policies that specifically address health disparities among indigenous minorities are necessary for long-term management. This includes arguing in favor of laws that protect the rights of indigenous peoples, advance equitable access to healthcare, and deal with socioeconomic determinants of health. It is also a part of the job to keep an eye on how policies are being implemented and how they are reducing health inequalities. When combined, these methods can help tribal minorities in Indochina manage health disparities by improving access to healthcare, addressing underlying issues, and empowering local people to actively participate in their health.

8 Lessons Learned for Future Direction to Manage Health Equity Among Tribal Minority Groups in Indochina

To manage health equity among tribal minority groups in Indochina, several lessons can be learned from previous experiences and initiatives. These lessons can guide future directions and strategies to address health disparities. Here are some key lessons. First and foremost, it's important to acknowledge and honor the cultural beliefs, customs, and languages of tribal minority groups. Involve community members and leaders in the planning and execution of healthcare interventions to guarantee cultural sensitivity and relevance. Secondly, indigenous minority communities must be involved in the planning, execution, and decision-making processes of programs. Promote community-driven initiatives and collaborations that enable individuals to assume responsibility for their health and welfare. Third, to increase access to healthcare services for tribal minority groups living in remote and neglected areas, it should develop and fortify the healthcare infrastructure in those areas. Primary healthcare centers, mobile clinics, and telemedicine services are being built to assist impoverished communities.

Fourth, it is crucial to develop targeted health education programs that tackle the particular needs and challenges faced by communities of indigenous minorities. Encourage the early detection of diseases, healthy lifestyle choices, and preventative interventions. Use culturally appropriate communication channels and materials to improve health literacy. Fifth, it ought to take into account resolving the underlying socioeconomic reasons for the disparity in health across tribal minorities. Implement initiatives for reducing poverty, providing vocational training, and generating revenue to decrease health disparities and boost economic opportunities. Sixth, while working with tribal minority populations, healthcare providers need to become more culturally competent. Teach cultural sensitivity and impart communication skills, awareness of tribal health requirements and customs, and cultural sensitivity training and education. Seventh, it needs to promote cooperation between various parties, such as governmental bodies, international partners, community-based groups, and non-governmental organizations (NGOs). Integrate expertise, resources, and information to develop thorough, long-term strategies for indigenous minority health equity. Eighth, data collection methods for indigenous minority health indicators should be improved. Regular monitoring and evaluation of operations is necessary to determine their impact and identify areas that require improvement. Finally, it must back practices and policies that give health equity a top priority and attend to the particular needs of communities of indigenous minorities. Involve legislators to push for their inclusion in national health agendas and to raise awareness of the health inequities that these groups face.

By putting these lessons into practice and employing a thorough, multi-sectoral strategy, the indigenous minority groups of Indochina may make significant headway in addressing health equity. To ensure these marginalized people have equitable and long-term health results, empowerment, community involvement, and cultural awareness should be given top priority going forward.

9 Centering Strengths of Traditional Knowledge and Culturally Sensitive Interventions

In Southeast Asia, health disparities and unequal access to healthcare services are common problems for indigenous peoples and ethnic minorities. Providing wellness interventions to these communities, however, can depend critically on identifying and highlighting the benefits of traditional knowledge and culturally appropriate techniques. Research on the value of incorporating traditional knowledge and culturally sensitive methods into healthcare delivery for ethnic minorities and indigenous peoples in Southeast Asia is crucial [11, 12].

Indigenous peoples' wisdom and rituals form the basis of traditional knowledge, which offers important insights into holistic health and well-being [13]. This information includes ancient medical practices, herbal remedies, spiritual practices, and cultural customs that have been passed down through the ages [11–13]. The integration of traditional knowledge into healthcare practices can facilitate the provision of more comprehensive and culturally sensitive care that honors the unique needs and values of these communities. Culturally sensitive methods are needed to address the healthcare requirements of ethnic minorities and indigenous peoples. To guarantee that healthcare services are provided in line with the cultural context of the community, these interventions are meant to be respectful of cultural values, norms, and traditions [13]. The cultural backgrounds of these people should be acknowledged and respected by healthcare professionals to foster better communication, increase trust, and promote involvement in healthcare decision-making.

Traditional wisdom and culturally aware interventions have the advantage of emphasizing overall wellness as opposed to just addressing physical problems. Ethnic minorities and indigenous peoples frequently possess a profound awareness of the connections between mental, emotional, spiritual, and physical health. Traditional healing methods and cultural customs can be included in healthcare interventions to suit the various needs of these communities and attain a more holistic approach to wellness.

10 Strategies to Address Health Inequity

Ensuring that everyone has access to high-quality healthcare services requires strengthening healthcare systems. Bridging the gap in healthcare inequalities can be achieved in part by increasing access to medical personnel and facilities in distant places. To do this, telemedicine services or mobile clinics can be established to serve people in underserved or rural areas [9, 10]. The implementation of community-based healthcare programs can aid in addressing the particular healthcare requirements of particular demographics [11, 12]. Reducing health disparities and boosting general well-being requires addressing social determinants of health. Reducing poverty and providing minority groups with educational opportunities are essential for enhancing health outcomes. Social and economic inequality can be reduced, giving people better access to opportunities and resources that improve their health. Encouraging social inclusion and battling discrimination can also improve mental and physical health [11, 12].

Enhancing workplace safety and environmental quality is essential for advancing public health. Environmental elements that can affect general health include exposure to chemicals, the quality of the air and water, and the availability of green spaces [11, 12]. We can lower the risk of illness and encourage a healthy living environment for everyone by addressing environmental variables. Preventing diseases and injuries at work is another reason why workplace safety is crucial. Workers' health and wellbeing can be safeguarded by providing appropriate training and ensuring safe working conditions. Individual healthcare professionals and clinics have an ethical responsibility to provide an environment that is openly affirming and culturally competent to assist address these discrepancies on a patient-by-patient basis, even as they advocate for systemic change [34].

To advance health equity and enhance general well-being, it is imperative that social determinants of health be addressed and healthcare systems be strengthened. To address the underlying causes of health disparities and advance health equity for all, policymakers, healthcare professionals, and community members must collaborate. We cannot attain a more equal and healthful future for everyone unless we work together.

11 Conclusion

This chapter concludes by addressing the significant challenges and problems of health equity for the indigenous minorities of Indochina. It emphasizes that, in comparison to the general population, there are notable disparities in healthcare access for specific ethnic groups, which lead to poorer health outcomes. The chapter looks at a variety of factors, including discrimination, poor infrastructure, remote location, and obstacles to culture. It also examines the unique health problems that tribal minorities face, including a higher risk of infectious diseases, undernourishment, and limited access to essential medical care. The chapter discusses the need to implement initiatives and programs that will fortify the healthcare system, increase service accessibility, improve the cultural competence of healthcare professionals, and empower Native Americans. It highlights how crucial collaboration is for long-term progress in health equity between governments, non-governmental organizations, and tribal peoples. To create a more equitable and inclusive healthcare system that supports the well-being and dignity of every person, the chapter concludes by highlighting the significance of putting comprehensive and inclusive measures in place to address the unique needs and rights of tribal minorities.

In summary, the Indochina area has formidable obstacles in tackling health disparities, especially for marginalized groups like ethnic minorities and rural people. There are differences in health and well-being due to restricted access to healthcare services in various sections of the region. These disenfranchised groups lack resources and infrastructure, and they also lack knowledge and instruction about health-related issues. Investing more in healthcare infrastructure, making healthcare services more accessible in rural locations, and offering ethnic minority groups healthcare services that are responsive to their cultural needs are some possible answers to these problems. Enhancing health outcomes and lowering disparities can also be achieved by supporting health education and awareness initiatives in these communities. To promote fair development and the wellness of all citizens in the region, health inequalities in Indochina must be given top priority.

In addition to enhancing people's general health and well-being in Indochina, tackling health inequities can help advance social cohesion and economic growth. For instance, Vietnam has significantly improved maternal health outcomes and decreased child mortality rates by implementing focused healthcare initiatives for rural and minority groups. Indochina may work toward building a more equitable and prosperous future for all of its citizens by prioritizing the needs of marginalized people and by continuing to invest in healthcare infrastructure and services.

References

1. Liburd LC, Hall JE, Mpofu JJ, Williams SM, Bouye K, Penman-Aguilar A (2020) Addressing health equity in public health practice: frameworks, promising strategies, and measurement considerations. Annu Rev Public Health 41:417–432
2. Hahn RA, Truman BI (2015) Education improves public health and promotes health equity. Int J Health Serv 45(4):657–678
3. Ramazanu S, Ang E, Mokhtar IA, Cahoon J, Del Pino S, Gomez S (2022) Centering strengths of traditional knowledge and culturally sensitive interventions: Strategizing provision of prediabetes wellness interventions for indigenous peoples and ethnic minorities in the United States and across Southeast Asia. J Diabetes 14(11):783–785
4. Bhatt DC, Neog R (1997) Health and development concerns of the north-east. Health Millions 23(3):3
5. Highland People Development Center (in Thai). Avilable online at http://www.mhsdc.org/
6. Målqvist M, Hoa DT, Liem NT, Thorson A, Thomsen S (2013) Ethnic minority health in Vietnam: a review exposing horizontal inequity. Glob Health Action 6:1–19
7. Nepote J (1976) In the golden triangle with a handful of dollars. Bull Narc 28(1):1–8
8. Lyttleton C, Cohen PT (2003) Harm reduction and alternative development in the golden triangle. Drug Alcohol Rev 22(1):83–91
9. Malaria. Available online at https://www.who.int/news-room/fact-sheets/detail/malaria. Access 15 Mar 204

10. Li JHLJ, Zhen YX (2012) Analysis on results of malaria surveillance in border areas of Guangxi Zhuang Autonomous Region during 2001–2010. Int J Med Parasit Dis 39:218–222
11. Gilmour B, Alene KA, Clements A (2022) The prevalence of tuberculosis and malaria in minority indigenous populations of South-East Asia and the Western Pacific Region: a systematic review and meta-analysis. Pathog Glob Health 116(4):201–219
12. Gilmour B, Alene KA, Clarke NE, Clements ACA (2021) The prevalence of tuberculosis, malaria and soil-transmitted helminth infection in minority indigenous people of Southeast Asia and the Western Pacific: protocol for a systematic review and meta-analysis. Syst Rev 10(1):203
13. Suswardany DL, Sibbritt DW, Supardi S, Chang S, Adams J (2015) A critical review of traditional medicine and traditional healer use for malaria and among people in malaria-endemic areas: contemporary research in low to middle-income Asia-Pacific countries. Malar J 14:98
14. Global Tuberculosis Report 2023. Malaria. Available online at https://www.who.int/teams/global-tuberculosis-programme/tb-reports/global-tuberculosis-report-2023. Access 15 Mar 204
15. Nguyen HV, Tiemersma EW, Nguyen HB, Cobelens FGJ, Finlay A, Glaziou P, Dao CH, Mirtskhulava V, Nguyen HV, Pham HTT, Khieu NTT, de Haas P, Do NH, Nguyen PD, Cung CV, Nguyen NV (2020) The second national tuberculosis prevalence survey in Vietnam. PLoS ONE 15(4):e0232142
16. Matsumoto-Takahashi ELA, Tongol-Rivera P, Villacorte EA, Angluben RU, Jimba M, Kano S (2021) Factors determining job satisfaction of malaria community microscopists in Palawan, the Philippines: a cross-sectional mixed-methods study. Int Health 13(1):22–29
17. GENIN (1947) Tuberculosis in Indochina. J Med Bord 124(4):174
18. Ong CC, Teo LLS (2022) Endemic thoracic infections in Southeast Asia. Radiol Clin North Am 60(3):445–459
19. Sultana ZZ, Hoque FU, Beyene J, Akhlak-Ul-Islam M, Khan MHR, Ahmed S, Hawlader DH, Hossain A (2021) HIV infection and multidrug resistant tuberculosis: a systematic review and meta-analysis. BMC Infect Dis 21(1):51
20. Mungmunpuntipantip R, Wiwanitkit V (2020) Expected coverage rate of tuberculosis screening among risk populations: a note from a rural community in the golden triangle area, Indochina. Popul Health Manag 23(4):336
21. Mastro TD, Coninx R (1988) The management of tuberculosis in refugees along the Thai-Kampuchean border. Tubercle 69(2):95–103
22. HIV statistics, globally and by WHO region, 2023. Available online at https://cdn.who.int/media/docs/default-source/hq-hiv-hepatitis-and-stis-library/j0294-who-hiv-epi-factsheet-v7.pdf. Access 15 Mar 204
23. HIV rates by countries. Available online at https://wisevoter.com/country-rankings/hiv-rates-by-country/. Access 15 Mar 204
24. Incidence rate of HIV per 1,000 population aged between 15 and 49 in Vietnam from 2010 to 2021. Available online at https://www.statista.com/statistics/1107580/vietnam-hiv-incidence-rate/. Access 15 Mar 204
25. Trotter AB, Hong SY, Srikantiah P, Abeyewickreme I, Bertagnolio S, Jordan MR (2013) Systematic review of HIV drug resistance in Southeast Asia. AIDS Rev 15(3):162–170
26. Apidechkul T (2016) A 20-year retrospective cohort study of TB infection among the Hill-tribe HIV/AIDS populations, Thailand. BMC Infect Dis 16:72
27. Apidechkul T (2019) Sexual behaviors and seroprevalence of HIV, HBV, and HCV among hill tribe youths of Northern Thailand. BMC Public Health 19(1):1101
28. Mandal P, Mittal S, Chadda RK (2023) Polysubstance use in South Asia. Curr Opin Psychiatry 36(4):269–276
29. Suwanwela C, Kanchanahuta S, Onthuam Y (1979) Hill tribe opium addicts: a retrospective study of 1,382 patients. Bull Narc 31(1):23–40
30. Akter S, Davies K, Rich JL, Inder KJ (2020) Barriers to accessing maternal health care services in the Chittagong Hill Tracts, Bangladesh: a qualitative descriptive study of Indigenous women's experiences. PLoS ONE 15(8):e0237002

31. Roesler A, Smithers LG, Winichagoon P, Wangpakapattanawong P, Moore V (2018) Health workers' and villagers' perceptions of young child health, growth monitoring, and the role of the health system in remote Thailand. Food Nutr Bull 39(4):536–548
32. Hill K, Ward P, Grace BS, Gleadle J (2017) Social disparities in the prevalence of diabetes in Australia and in the development of end stage renal disease due to diabetes for Aboriginal and Torres Strait Islanders in Australia and Maori and Pacific Islanders in New Zealand. BMC Public Health 17(1):802
33. Moonpanane K, Pitchalard K, Thepsaw J, Singkhorn O, Potjanamart C (2022) Healthcare service utilization of hill tribe children in underserved communities in Thailand: Barriers to access. BMC Health Serv Res 22(1):1114
34. Lund EM, Burgess CM (2021) Sexual and gender minority health care disparities: barriers to care and strategies to bridge the gap. Prim Care 48(2):179–189

14 Tackling Healthcare Disparities in Severe Mental Disorders: Integrating Physical Activity Therapies and Professionals into Mental Health with the Psychiactive Project

López-Moral Álvaro, López-Sánchez Camilo, Borrueco-Sánchez Jesús, Munguía-Izquierdo Diego, and Bueno-Antequera Javier

Abstract

This chapter addresses the issue of healthcare disparities among individuals with severe mental disorders (SMD) by focusing on the integration of physical activity therapies as an adjunctive treatment facilitated by physical health professionals within the current mental health paradigm. Emphasis is placed on the compelling need for a holistic and multidisciplinary approach, considering that prevailing mental health services often overlook physical exercise as a pivotal element in both the prevention and treatment of this escalating concern. To achieve this goal, an enhanced understanding of how such interventions can be meticulously designed and effectively implemented in "real-world" conditions is imperative to optimize patient adherence. This chapter reviews selected case studies that have successfully implemented physical activity programs within mental health services, thereby enhancing the overall health and quality of life of people with SMD. Moreover, the chapter explores the existing

L.-M. Álvaro · L.-S. Camilo · B.-S. Jesús · M.-I. Diego · B.-A. Javier
Physical Performance and Sports Research Center, Universidad Pablo de Olavide, Sevilla, Spain

L.-M. Álvaro
Department of Physical Education and Sport, Universidad de Sevilla, Seville, Spain

M.-I. Diego · B.-A. Javier
Research Group in Development Movimiento Humano, Universidad de Zaragoza, Zaragoza, Spain

M.-I. Diego (✉)
Department of Sport and Computer Science, Universidad Pablo de Olavide, Carretera Utrera Km. 1, s/n, 41013 Sevilla, Spain
e-mail: dmunizq@upo.es

B. W. K. Son (ed.), *The Landscape of Global Health Inequity*, Integrated Science 22,
https://doi.org/10.1007/978-3-031-60502-4_14

barriers impeding the effective implementation of this innovative approach in mental health, offering insightful strategies to overcome these obstacles. Taking this evidence together and undertaking a review of the updated literature, this chapter provides a robust foundational framework for comprehending the significance and feasibility of integrating physical activity therapies into the comprehensive treatment paradigm for SMD.

Keywords

Lifestyle psychiatry • Multidisciplinary approach • Adherence • Dropout • Case studies • Barriers • Facilitators • Behavior change techniques

1 Introduction

People with severe mental disorders (SMD), such as schizophrenia, depression, anxiety, and bipolar disorder, are two to three times more likely to die than healthy people in every age and gender range, this translates into a reduction in life expectancy of about 10–20 years [1, 2]. This phenomenon is attributed to the markedly elevated prevalence of various physical diseases, such as cardiovascular disease, osteoporosis, decreased bone mineral density, impaired lung function, obesity, and diabetes [3]. Most patients with a psychotic disorder have many of the symptoms that comprise metabolic syndrome defined as having three of its five components (high triglyceride and glucose levels, low high-density lipoprotein levels, elevated blood pressure, and waist circumference) [4], which is a major contributor to the declining health and longevity of individuals with SMD. This population is at high risk of getting these symptoms, which is critical for their physical health and makes them more vulnerable to cardiovascular diseases. The simultaneous presence of these adverse medical conditions not only negatively impacts the quality of life of affected individuals but also leads to a substantial increase in vulnerability to critical health problems, underscoring the imperative need for intervention strategies and comprehensive care to improve well-being and life expectancy in this specific population.

On the other hand, people with SMD exhibit unhealthy lifestyle habits, most notably a sedentary lifestyle, conceptualized as "any activity characterized by energy expenditure $\leq$ 1.5 METs while sitting or lying down" [5], with physical activity levels considerably lower than standard recommendations. Furthermore, this population tends to experience sleep-related problems, constituting an additional risk factor for the development of the aforementioned diseases, although the complexity of this relationship requires further scrutiny through more comprehensive research [6]. Along with lifestyle, smoking is another major problem in people with SMD that has a direct negative impact on cardiovascular health [7]. Tobacco use is one of the main reasons individuals with SMD experience a decrease in life expectancy. In fact, among the same population, it has been observed that individuals who do not smoke have a longer life expectancy than those who do [8].

The mortality rate is the most effective objective indicator for assessing the results of treatment and recovery from a disease. From 1952 to the present day, health policies have relied on pharmacological treatment supported by psychotherapeutic interventions to combat the symptoms of SMD. However, the pharmacological use of antipsychotics is closely related to a higher prevalence of hypertension, obesity, diabetes, high cholesterol, dyslipidemia, and other factors that increase the risk of death from cardiovascular disease [9, 10]. This may explain why, despite these treatment efforts, the current growth trend in this population has not been reduced, creating a global problem at the economic, social, and family levels.

2 Physical Activity Therapies and Professionals in Mental Healthcare

2.1 Benefits of Integrating Physical Activity Therapies and Physical Activity Professionals in Mental Healthcare

In recent decades, the effects of various physical exercise programs have been studied in individuals with SMD to identify benefits at different levels [11]. Accumulating evidence supports physical exercise as an effective and safe complementary therapeutic intervention capable of improving the quality of life, clinical symptoms, global functioning, and mood in this population [12, 13]. Increased levels of physical activity in this group of individuals lead to improvements in cardiometabolic health, cardiorespiratory fitness, functional capacity, quality of life, psychiatric symptoms, and cognitive function in people with SMD [11]. Implementation of a physical exercise program in individuals with SMD, to improve cardiorespiratory fitness and strength, is also associated with improvements in disease symptomatology, functional disability, and cognition [14, 15]. In addition, participation in a physical exercise program by this population appears to be linked to benefits in reducing inflammation and increasing oxidative stress while promoting self-esteem, social support, and self-efficacy, which are often significantly impaired in these individuals [16].

Several studies have conclusively indicated that physical exercise is an effective and safe coadjuvant therapy to improve physical health outcomes such as cardiorespiratory fitness and muscular strength in SMD [17]. These aspects include cardiovascular and respiratory health, quality of life, disease symptoms, obesity, and an individual's physical condition, addressing both muscular strength and aerobic capacity. It is important to emphasize that physical exercise is not intended to replace any of the therapeutic interventions or treatments carried out in the care of people with mental health problems but is positioned as a beneficial complement when incorporated alongside other therapeutic modalities in the comprehensive management of this population. Some of the most recent studies with outstanding results are presented in Table 1, selected for their relevance and compliance with the FITT (Frequency, Intensity, Time, and Type) standards.

2.2 Integrated Approach with Practitioners: Lifestyle Psychiatry

It is imperative to adopt a holistic approach to the treatment of mental health problems, incorporating consideration of physical activity, sleep, smoking, and nutritional patterns as complementary components of notable relevance in influencing the risk, development, and therapeutic outcomes of mental disorders [8]. These variables, as previously mentioned, are significantly compromised in the affected population, thus justifying the inclusion of these factors in the treatment plan to improve the quality of life and physical health of these individuals.

The harmful habit of daily tobacco use among people with SMD is associated with an increased risk of psychotic disorders and earlier onset of mental illness, constituting a significant and modifiable risk factor [18]. Similarly, sleep quality, in which interventions appear to reduce depressive symptoms, especially in individuals with mental disorders, has emerged as another crucial aspect [19]. With regard to nutrition, despite less evidence, an association between dietary risk, cardiometabolic diseases, and mental disorders has been established, justifying the consideration of dietary factors in the context of multidisciplinary healthcare for people with mental illness [20].

In addition, physical exercise plays a prominent role, as previously noted, which brings substantial benefits to people with mental health problems, including being a protective factor in reducing the risk of certain mental disorders [8]. Furthermore, physical exercise has been shown to play an effective complementary role in the treatment of depression, anxiety, and other mental disorders. To effectively integrate physical exercise into mental health treatment, it must be planned and supervised by exercise professionals, such as physical-sport educators, who have the necessary training to conduct safe and effective training for this population.

2.3 The Problem of Adherence and Dropout Rates

Individuals with SMD present difficulties in terms of adherence to physical exercise, exhibiting notably high dropout rates during physical activity. Based on the meta-analysis conducted by Vancampfort et al. [21], a dropout rate of 26.7% was identified; that is, more than one in four individuals who started a physical exercise program discontinued it. These high figures are linked to various factors that limit the benefits or positive effects of these interventions on the health of this population, which are closely related to the barriers faced by this population towards the practice of physical exercise. Lack of continuity in training programs deprives these individuals of reaping the full benefits associated with physical exercise, as

Table 1 Recent studies on physical exercise in SMD

Author	Aim	Population	Intervention	Desing	Intensity	Supervision	Results
Chau et al. [74]	Investigate both short and long-term effects of a structured physical rehabilitation program	Mayor depressive disorder (42 vs. 42 control)	ME (AE + RE)	3 d/wk of 60 min over 12 weeks	50–70% HRmax	Yes	The intervention group improved cardiopulmonary function, depressive symptoms, pain, body composition, muscle strength, and flexibility
Imboden et al. [75]	Evaluate the effect of AE on physiological outcomes and cognitive performance	Depression (22 vs. 20 control)	AE (cycling)	3 d/wk of 45 min over 6 weeks	60–75% HRmax	Yes	Improvements in symptom severity, mental toughness, physical self-description, endurance score, cognitive flexibility, and BMI

(continued)

Table 1 (continued)

Author	Aim	Population	Intervention	Desing	Intensity	Supervision	Results
Curcic et al. [76]	How prescribed physical activity influences their cardio-respiratory fitness and symptoms of schizophrenia	Schizophrenia (25 vs. 23 control)	AE (walking)	5 d/wk of 45 min over 12 weeks	65–75% HRmax	Yes	Exercise group showed a significant increase of VO2max compared to the control group. Significant differences were also observed on PANSS general psychopathology subscale and on PANSS total score
Şenormancı et al.[77]	Investigate the effects of exercise on psychotic symptoms, depression, functionality, insight, and resilience in patients with schizophrenia	Schizophrenia (20 vs. 19 control)	RE	2 d/wk of 60 min over 12 weeks	65–75% HRmax	–	Exercise combined with TAU may be effective in increasing awareness of illness and alleviating depression in chronic schizophrenia

Note AE Aerobic Exercise, *RE* Resistant Exercise, *ME* Multimodal Exercise, *d/wk* days/week, *HR max* maximum Heart Rate, *PANSS* Positive and Negative Syndrome Scale for Schizophrenia, *BMI* Body Mass Index discussed above. Therefore, the presence of a physical sports educator is crucial to increasing adherence and reducing the dropout rate.

The scientific literature offers several studies that compare adherence and dropout of training programs among people with SMD when the intervention is led by physical sports educators or by individuals who are not qualified or trained to perform such interventions. The results indicate that exercise is well tolerated by those who participate in programs led by qualified professionals, showing a significantly lower dropout rate than interventions led by unqualified personnel. This finding underlines the need to incorporate physical sports educators into mental health facilities [21–23].

Effective collaboration and coordination between health centers and physical sports educators is essential for the successful implementation of exercise intervention programs. Since patients often follow the recommendations of the health center, 65% of them reported having started exercise programs thanks to the guidance or insistence of a professional linked to the center [24]. The involvement and support of the center's staff towards patients, urging them to start and support them, play a crucial role in overcoming fears associated with physical exercise, as pointed out by Curbelo et al. [25].

Additionally, the difficulty experienced by this population in socializing is addressed by scientific evidence, which suggests that physical exercise facilitates the creation of stronger bonds, thus increasing adherence [26]. Given that people with SMD attach great importance to companionship during physical activity, as well as to the relationships that are established between the patients themselves and the professionals who train them, as well as with the physical sports educators present, group cohesion and a sense of belonging are generated that contribute to increased adherence to exercise programs, generally characterized by a low rate of continuity [27].

3 Case Studies: Integrating Physical Exercise into Mental Health

As discussed above, people with SMD have unhealthy lifestyle habits, including low levels of physical activity, prolonged periods of sedentary lifestyles, smoking, poor dietary patterns, and suboptimal sleep quality [8]. These behaviors are health risk factors, but they are modifiable, and exercise and physical activity play key roles in modifying them. In 2019, the Lancet Commission on Psychiatry [28] highlighted the importance of incorporating tools aimed at improving the lifestyles of people with SMD, including the Lifestyle Psychiatry approach. In addition, various institutions of political and health relevance, such as NICE, APA, and the UK College of Psychiatrists, have integrated guidelines on healthy lifestyle habits, including physical exercise, into health promotion and quality-of-life policies. The implementation of these new policies has increased physical exercise interventions and programs in mental health facilities worldwide. The following are five success stories from recent years where physical exercise programs have been successfully implemented in these types of centers or hospitals.

3.1 Challenges in Implementing an Exercise Intervention Within Residential Psychiatric Care: A Mixed-Methods Study

The study conducted by Firth et al. [29] took place in Manchester, UK, to analyze the challenges associated with implementing a physical exercise intervention in a residential psychiatric care setting. In total, 14 participants were selected for the intervention, which lasted for 10 weeks, with two exercise sessions per week. One session consisted of a traditional workout, while the other was a self-chosen physical activity, aiming for a total of 90 min of exercise per week. Each session included 10 min of warm-up, 45 min of circuit training, and 5 min of cool-down. The circuits consisted of 8–16 stations with aerobic and strength exercises selected to improve cardiovascular fitness, muscular strength, and functional capacity. All the sessions were supervised by an exercise professional. The key findings of the study supported the implementation of the physical exercise program at the center:

- There was a 12.2% reduction in Positive and Negative Syndrome Scale for Schizophrenia (PANSS) symptom scores after the intervention.
- A 21% decrease in negative PANSS symptoms was observed.
- A significant association was identified between increased physical activity and decreased psychiatric symptoms (PANSS).
- A 6.5% improvement in socio-occupational functioning was achieved.
- A reduction in body weight was recorded.

In addition to the quantitative data, the qualitative analysis revealed significant aspects. Participants indicated that they engaged in physical exercise mainly to "increase energy and/or fitness" (92.9%), "have fun" (78.6%), and "stop thinking" (64.3%), among other incentives. However, some limitations of the intervention were identified, such as lower than expected recruitment rates of approximately 30%. This suggests a low interest of people in participating in physical activity at that time, indicating a lack of motivation for this therapy. In addition, the level of physical activity was measured subjectively through a recording sheet, with no objective monitoring of activity behaviors, which may affect the reliability of the data obtained.

3.2 Evaluating the Feasibility of a Pilot Exercise Intervention Implemented Within a Residential Rehabilitation Unit for People with Severe Mental Illness: GO HEART (Group Occupational Health Exercise and Rehabilitation Treatment)

Korman et al. [30] evaluated the feasibility of implementing a physical exercise program in a rehabilitation unit for people with SMD. The intervention was developed in the rehabilitation unit of the Metro South Addiction Health Service in Brisbane, Australia, and involved a total of 13 subjects selected for the physical exercise program. The implementation of this intervention in this specific population was based on the following four key considerations [31]: assessment of cardiovascular risk, cardiorespiratory function assessment, programming of the appropriate intensity level, and strategies to improve motivation and adherence.

The duration of the program was 10 weeks, with sessions three times per week to maximize fitness improvement [32]. The program consisted of a station-based circuit based on 10–15 aerobic and strength exercises with a total duration of 45 min. The circuit varied slightly each week, allowing the participants to adjust their loads according to their abilities and preferences. The training was supervised by final-year physical exercise students. The intervention incorporated evidence-supported motivational strategies to increase the participants' adherence to training. Coaches and mental health professionals provided support and encouragement during the intervention, helping participants set realistic goals and address individual preferences in planning the exercise program. Team meetings were held between coaches and psychiatrists to discuss and resolve potential barriers to participants' participation in the program.

The results of the program were remarkable, with a recruitment rate of 81% and an adherence rate of 77%, which is high considering the clinical population [22, 33]. Of the initial 13 participants, 10 completed the program, and the average session attendance was 78%. These adherence rates are comparable to those of psychological interventions for people with schizophrenia and depression [34, 35]. The flexibility of the program was highlighted by inviting participants with morning sedation to perform recovery sessions at another time of the day and adapting circuits for those who regularly skipped sessions. Significant improvements were observed in relation to the health variables analyzed:

- Significant increase in functional capacity, as measured by the increase in distance covered in the 6-min test (6MWT).
- Increase in minutes of physical activity, with 80% of participants increasing their exercise volume by more than 60 min.
- Significant reduction in psychiatric and negative symptoms of illness, with 70% of participants decreasing the PANSS score by more than 10 points.
- Significant reduction in intrinsic regulation, as measured by BREQ-2.

Despite these encouraging results, intervention limitations were identified. Although the staff in the rehabilitation unit received pre-implementation training, it was insufficient to change the culture of the staff and to adequately prioritize the physical exercise program over other daily responsibilities. In addition, the participants were young and had no functional problems or comorbidities, which limited the generalizability of the findings.

3.3 Participation in and Satisfaction with an Exercise Program for Inpatient Mental Health Consumers

Stanton et al. [36] conducted an analysis of participation and satisfaction in a physical exercise program for hospitalized people with SMD. The study was conducted in a private hospital in Queensland, Australia. The study sample consisted of 37 participants who were exposed to five activities offered in the hospital setting: cognitive behavioral therapy, creative expression, relaxation, reflection/discussion, and physical exercise. Each of these activities is voluntary. The physical exercise group was supervised by a physical exercise specialist and consisted of a 50-min training session held twice a week. This session included aerobic and strength exercises, with 20 min of aerobic activity on a treadmill or cycloergometer at a moderate intensity according to the patient's preference, followed by 20 min of a strength circuit with bands and free weights, performing two to three sets and 15 repetitions. Exercise intensity was selected by the patient.

The results of this study indicated that physical exercise therapy received the most positive responses from patients. Of the 28 participants who responded, 16 (almost 60%) expressed that physical exercise therapy was excellent. In addition, participation in this therapy was high, considering that participation was voluntary. These findings suggest that individualized physical exercise in a hospital setting for patients with SMD is well accepted and can generate high satisfaction when implemented in an appropriate context and led by an exercise professional.

However, certain limitations were identified in the study, such as the possible tendency of the satisfaction scale to favor positive responses without specifically identifying which aspect of each therapy may have been responsible for the level of satisfaction. In addition, the results cannot be generalized, as other inpatients may be exposed to different therapies, making direct comparisons difficult.

3.4 Multidisciplinary Lifestyle-Enhancing Treatment for Inpatients with Severe Mental Illness (Multi-study): Effects on Physical Health, Psychotic Symptoms, Quality of Life and Functioning

In a study conducted by Deenik et al. [37], a Multidisciplinary Treatment Program to Improve Lifestyle in Hospitalized Patients (MULTI) with SMD was implemented. The program was developed in the Netherlands and aimed to assess changes in mental and physical health and functioning after 18 months compared to treatment as usual (TAU).

The intervention, which covered 18 months, involved 114 inpatients, of whom 65 underwent the MULTI program and 49 received TAU. The MULTI program consisted of a structured daily routine based on sports and work activities, attention focused on dietary habits, psychological education, and active participation of the center's staff. It was characterized by cultural change based on the principle of "change from within" through multidisciplinary collaboration. Measures ranging from activity behaviors to quality of life, symptoms, and psychosocial functioning, among other variables, were collected.

The results showed positive effects for most of the variables analyzed. A reduction in body weight, increase in moderate–vigorous physical activity, improvements in metabolic parameters, improved psychosocial functioning, and decrease in social problems were observed. In addition, a significant increase in the quality of life was observed. The analysis of this study revealed that the negative trend of deterioration in these patients could be stopped, some relevant parameters could even be positively reversed, and no negative effects were observed. Likewise, the MULTI program proved to be superior to TAU, positioning itself as a viable alternative and solution compared to the current resources.

3.5 Keeping the Body in Mind: A Qualitative Analysis of the Experiences of People with First-Episode Psychosis Participating in a Lifestyle Intervention Program

Watkins et al. [38] implemented the Keeping the Body in Mind (KBIM) program at the Bondi Centre in Sydney, Australia. The intervention aimed to explore the personal experiences of individuals with SMD who had just experienced their first psychotic episode and participated in a healthy lifestyle program as part of their episode treatment.

Eleven participants, aged 18–25 years and diagnosed with schizophrenia, depression, or bipolar affective disorder, underwent an intervention lasting 12 weeks. This program aimed to prevent weight gain and metabolic health deterioration. Three health professionals collaborated in this initiative: a dietitian, a clinical nurse, and an exercise physiologist. Each professional offered specific services to broaden their perspective on health improvement. The exercise physiologist provided individualized intervention plans, and the participants had access to a supervised gym and participated in weekly sports activities. At the end of the intervention, interviews were conducted with participants about various aspects of the healthy lifestyle program and their opinions on KBIM.

Analysis of the data revealed four main themes across the interviews: the role of physical health in mental health recovery; the relationship and interaction with professionals, understood as support, positive change, and encouragement from staff; the value of peer interaction and a sense of belonging to a community; and gradual change towards a sustainable healthy lifestyle.

The KBIM model has demonstrated that the effects of lifestyle interventions span a broader spectrum than those of simply improving physical health. These types of programs are highly valued in the recovery of mental health among young people with psychosis and are considered a priority in services for these types of mental health problems. Despite the positive results, some limitations should be considered, such as the inability to generalize the results because the participants were young people and the presence of a member of the KBIM program in the interviews, which could have biased the participants' opinions toward positive comments.

3.6 Key Points of the Implementation of Physical Exercise Programs in Mental Health Services

In general, the challenges and proposals for future research in mental health centers that incorporate physical exercise are derived from the key points extracted from the studies analyzed. The need to individualize each participant so that the subject feels comfortable and safe within the exercise program, thus achieving his or her goals and committing to maintaining continuity of training, which may result in reduced dropout rates, is highlighted. Given the importance of intrinsic motivation in maintaining exercise regularity in individuals with SMD, developing effective exercise interventions that are considered fun and maintain interest over time becomes relevant and should be an important consideration.

Additionally, it seems essential that mental health professionals collaborate closely with exercise professionals to appropriately implement physical exercise programs in mental health facilities. Meetings to address doubts, solve problems, and demystify misconceptions about exercise should be considered a priority to change the culture within the centers and enhance interventions aimed at improving the quality of life of patients.

To achieve a clinical impact on physical health outcomes, engagement in physical exercise must extend beyond brief interventions. Future trials of exercise interventions should extend beyond 24 weeks, with follow-up periods exceeding one year. In addition, multidisciplinary interventions that integrate various lifestyle therapies, such as physical exercise combined with nutritional interventions, represent a broader perspective on the integration of healthy lifestyle habits into mental health treatment, contributing to an improved quality of life in people with SMD [39].

Finally, the importance of supervision by qualified exercise science personnel was emphasized. Given the barriers faced by people with SMD, ongoing supervision and encouragement are critical for maintaining motivation in exercise performance [40, 41]. In addition, the training of trainers specializing in mental health should be a priority in selecting competent and appropriate staff to supervise sessions for people with SMD. The key points of integrating exercise into mental health services are summarized in Fig. 1.

4 Challenges and Future Perspectives

4.1 Barriers to Physical Exercise

Despite the inherent benefits of exercise and positive outcomes obtained through various interventions, individuals with SMD experience a variety of barriers. These barriers, conceptualized as psychological, physiological, and socioecological conditions, act as obstacles that decrease participation in physical activities [41, 42]. This limitation affects the effectiveness of adjuvant therapies in improving the mental health of this population. In this context, one of the preeminent studies was that of Firth et al. [41], who conducted a comprehensive review and meta-analysis to categorize the barriers faced by individuals with SMD. On a physical level, they highlight factors such as fatigue and lack of energy, whereas on a psychological level, the symptomatology of the illness, such as the presence of anxiety or depression, emerges as a critical component. From a social perspective, this study identified the lack of support and the absence of peers as elements that discourage participation in physical activity.

This conclusion is in line with findings from other studies in SMD populations, where the main barriers included fatigue, lack of social support (lack of companionship for exercise), lack of motivation, and adverse weather conditions [43]. Likewise, Chapman et al. [44] highlighted poor physical and mental health, along with the scarcity of financial resources, as preeminent impediments. In the specific case of schizophrenia, the review by Bernard et al. [45] also identifies fatigue, exercise inexperience, medication side effects, schizophrenia symptoms, and high levels of anxiety as significant barriers to the adoption of physical activity, a position shared by other authors [46, 47]. When examining people with schizophrenia, Soundy et al. [48] highlighted a lack of motivation, an aspect observed in other studies, and emphasized the lack of priority given to this resource by other health professionals. Additionally, mental health problems and their symptomatology have been identified as barriers by other authors [46, 49]. The literature also highlights less-explored barriers, such as negative feelings, unfavorable experiences with exercise, distortions in body perception, and misconceptions about physical activity [49]. Aspects such as lack of self-esteem, social anxiety related to physical activity, inadequate environments, and social isolation have emerged as elements that deserve attention, as they can also act as limiting barriers [47, 50].

As can be seen, the literature highlights the complexity of the barriers that hinder regular participation in physical exercise in individuals with SMD. Therefore, it is imperative to pay attention to psychological variables, such as lack of motivation or depressive symptomatology, as well as sociodemographic and physical variables, such as lack of companionship and fatigue.

4.2 Goals and Facilitators Towards Physical Exercise

Knowing the goals or objectives, conceptualized as patients' perceived reasons for increasing their participation in physical activity, could be crucial for designing and targeting specific interventions for these aspirations. A meta-analysis by Firth et al. [41] identified the main physical goals of people with SMD as improved physical fitness and increased energy, with 75% of participants agreeing with this purpose. Likewise, goals such as physical health stood out at 91%, improved physical appearance at 77%, weight loss at 83%, and strengthening at 72%. From a psychological perspective, the most pursued goals were improving overall mental health (80%), reducing stress levels (78%), greater mood control (81%), improving sleep patterns (72%), and seeking fun (54%).

Regarding possible gender differences, it was observed that men presented higher scores in goals such as improving image or seeking fun, while no significant differences were detected in goals such as improving physical fitness, socialization through exercise, or the desire to improve skills [51].

In addition, the consideration of facilitating aspects, understood as factors that reduce barriers and improve adherence to sessions [52], emerges as an aspect of special relevance. According to Soundy et al. [48], social support from professionals, promotion of fun, and support for autonomy are crucial elements in this context. These findings are consistent with a more recent systematic review by Méndez-Aguado et al. [52], which highlighted those outdoor activities, the inclusion of social components, and the possibility of adapting activities to individual needs can facilitate participation in physical exercise in this population. Likewise, other authors have emphasized the importance of professional advice as a determining factor in establishing exercise habits [43].

4.3 Motivation as a Key Factor

Not only is it of interest to identify barriers and goals, but paying attention to motivational variables on the part of physical educators could play a key role in expanding participation and adherence of individuals with SMD to physical exercise programs [53]. According to the self-determination theory [54], group managers should foster more self-determined motivations in individuals who compose such groups. This theoretical framework, embedded in social-cognitive theories, focuses on explaining behavior primarily in terms of the degree of voluntariness that a person perceives about his or her own behavior. Depending on this degree, people develop different types of motivation differentiated by the promotion of adaptive behaviors [54].

According to the degree of self-determination, a continuum is configured that begins with intrinsic motivation for intrinsic regulation, where people exercise for the mere enjoyment of the activity. Second, within extrinsic motivation, there is integrated regulation, which refers to exercise as part of daily routines and established values. Third, the extrinsic regulation motivation identified involves exercising because of knowledge of the benefits, even if the activity is not necessarily pleasurable. These three regulations constitute autonomous motivation and high-quality motivational regulations. Following the continuum, among the regulations to be avoided because of their lower motivational quality are introjected extrinsic regulatory motivation, which is associated with exercising because of feelings of guilt without a clear understanding of the activity, and external regulatory extrinsic motivation, where exercise is performed under external pressure or punishment. The risk inherent in these last two regulations lies in the fact that if the person persists in these motivational states, that is, exercising out of guilt and external pressures, it could evolve into demotivation, culminating in a lack of interest and eventual abandonment of the program [54]. The Self-Determination Continuum is illustrated in Fig. 2.

To foster high-quality motivational regulations, the theory postulates that physical educators should support, rather than frustrate, three basic psychological needs: autonomy, competence, and social relationships [55]. Perceived autonomy refers to a person's decision-making power over his or her own actions and opportunities for choice. Competence involves the perception of being able to perform assigned tasks satisfactorily. Finally, relationships relate to having quality social interactions where other people care and provide support [54].

In cross-sectional studies in the field of SMD, it was observed that higher levels of demotivation and external regulation were found in less physically active individuals, whereas identified and intrinsic regulations were lower in these individuals [56]. This finding is consistent with more recent studies in which autonomous motivation was significantly lower in subjects who did not exercise regularly [53]. Therefore, fostering more self-determined motivations seems to be essential to obtaining positive consequences at the behavioral level, such as greater adherence and persistence in practice, at the cognitive level, such as greater attention and concentration on the task, and at the affective level, such as greater enjoyment [55].

4.4 Behavior Change Techniques Are the Next Step

With this theoretical framework as a reference point, practical implementation is based on the use of behavior change techniques (BCTs) [57], that is concrete actions through which interventions that combine exercise and behavioral variables are carried out. Some of the most commonly used BCTs in the scientific literature in various populations include setting behavioral goals (e.g., meeting a daily step challenge), self-monitoring (e.g., teaching to use a step log or session attendance), instructions on how to perform tasks and exercises, providing social support (e.g., having peers to attend sessions), the use of feedback (e.g., reports reflecting progress), designing graded tasks, and using a reward system [58–64].

Several studies have evaluated the use of BCTs in physical activity programs for patients with SMD. Cody et al. [65], with a sample of 220 persons with major depressive disorder, observed that after the intervention, the group that received counseling with goal setting, plan review, and barrier resolution reduced their daily steps and moderate-to-vigorous physical activity less at the end of the intervention. Similarly, in a sample of 57 people with psychotic disorder [66], after 12 sessions of exercise combined with BCTs, the intervention group experienced a significant increase in total physical activity and moderate–light physical activity levels, as well as a decrease in positive illness symptoms. On the other hand, Seymour et al. [67], in a study of 95 people with mental illness, achieved increases in autonomous motivation and strength levels and decreases in distress. Similarly, Zeibig et al. [68] implemented a running program and BCTs that included goal setting, self-monitoring, and barrier resolution and achieved significant improvements in global symptomatology, depression, anxiety, sleep quality, and exercise quantity. This decrease in symptoms has also been observed in other studies [69, 70]. Additionally, interventions with BCTs seem to show good adherence and motivation [71, 72]

In light of these results, the incorporation of these group management techniques by physical educators in exercise interventions could be of interest to improve levels of adherence to the program and even encourage the acquisition of habits in people's daily lives. Once such continuity in the practice of physical exercise is achieved, individuals can experience improvements in health-related parameters, such as hip circumference, strength levels, or aerobic endurance, that is, anthropometry and physical fitness [74].

5 Conclusions

Most mental health hospitals, associations, foundations, and routine care centers do not currently incorporate physical exercise as part of the treatment or prevention of SMD. Limited factors such as inadequate training, lack of experience, and insufficient resources among professionals hinder the promotion of an active lifestyle among users. The implementation of these programs at the clinical level is met with resistance, resulting in patients experiencing insufficient professional support, impeding their adherence and compliance. Further trials are required to investigate the feasibility and sustainability of lifestyle interventions in mental health settings in this vulnerable population. To achieve this goal, it is crucial to comprehend the relationship between various barriers and facilitators of an intervention, the moderating behavior of diverse strategies, and the outcomes of the intervention. Ultimately, incorporating physical exercise as a tool for the treatment and prevention of SMD necessitates a cultural, social, and political shift that considers treatment from a multidisciplinary perspective and integrates physical sports educators as additional health agents.

Acknowledgements The authors would like to thank all the participants for their collaboration. We also thank the members of the Mental Health Day Center of the Hospital de Valme (Seville, Spain) and the Care Center of Portobelo FAISEM (Seville, Spain) who participated in the intervention for their efforts and willingness. This work was funded by Proyectos I+D+i 2020 (Ref. PID2020-118262RB-I00) from the Spanish Ministry of Science and Innovation and Universidad Pablo de

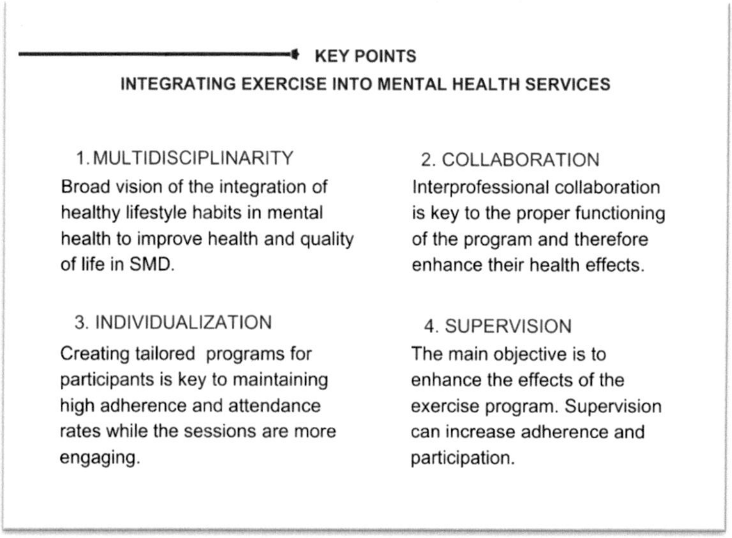

Fig. 1 Key points of integrating exercise into mental health services. *Note* SMD: Severe Mental Disorder. *Source* Own elaboration

Olavide (Ref. PPI2304). It was also supported by the CTS-948 Research Group, Universidad Pablo de Olavide, Junta de Andalucía, and FEDER Programme 2014-2020 (Ref. UPO-1262802). Universidad Pablo de Olavide supported ALM (Ref. PPI1803), JBS (Ref. PPI2104 and PPC2401), and CLS (Ref. PPI2306). JBA was supported by the Operational Programme FEDER Andalucía 2014-2020 (Ref. PAC2042) and by the Junta de Andalucía through the Consejería de Transformación Económica, Industria, Conocimiento y Universidades (Ref. POSTDOC_21_00059). The funders were not involved in the design of the study; the collection, analysis, and interpretation of the data; the writing of the report; or the decision to submit it for publication.

Appendix: Attachments

See Table 1 and Figs. 1, 2.

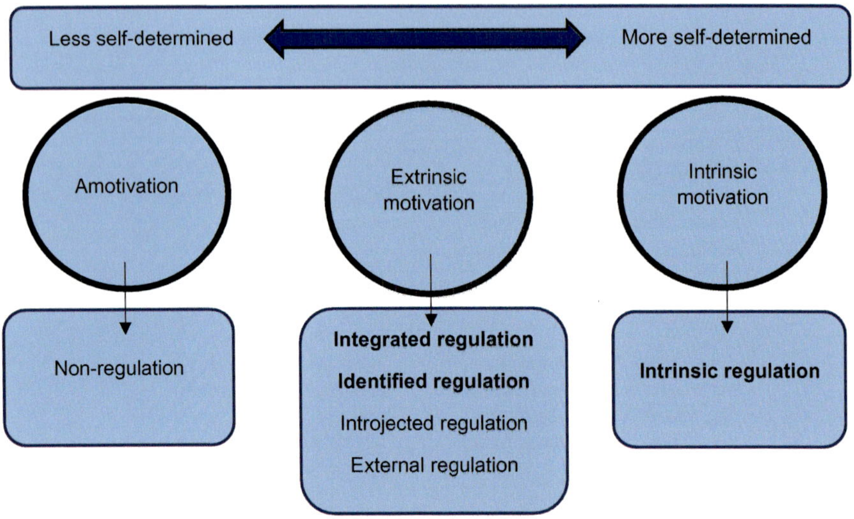

Fig. 2 Self-Determination Continuum. *Note* Motivational regulations marked in bold belong to autonomous motivation. *Source* Own elaboration based on Ryan and Deci [54]

References

1. Chan JKN, Tong CHY, Wong CSM, Chen EYH, Chang WC (2022) Life expectancy and years of potential life lost in bipolar disorder: systematic review and meta-analysis. Br J Psychiatry J Mental Sci 221(3):567–576. https://doi.org/10.1192/bjp.2022.19
2. Liu X, Wang D, Fan R, Wang R, Xiang H, Yang X, Liu Y (2022) Life expectancy and potential years of life lost for schizophrenia in western China. Psychiatry Res 308:114330. https://doi.org/10.1016/j.psychres.2021.114330
3. De Hert M, Correll CU, Bobes J, Cetkovich-Bakmas M, Cohen D, Asai I, Detraux J, Gautam S, Möller HJ, Ndetei DM, Newcomer JW, Uwakwe R, Leucht S (2011) Physical illness in patients with severe mental disorders. I. Prevalence, impact of medications and disparities in health care. World Psychiatry 10(1):52–77. https://doi.org/10.1002/j.2051-5545.2011.tb00014.x
4. Mazereel V, Detraux J, Vancampfort D, Van Winkel R, De Hert M (2020) Impact of psychotropic medication effects on obesity and the metabolic syndrome in people with serious mental illness. Front Endocrinol 11:573479. https://doi.org/10.3389/fendo.2020.573479
5. Sedentary Behaviour Research Network (2012) Letter to the editor: Standardized use of the terms "sedentary" and "sedentary behaviours." Appl Physiol Nutr Metab 37(3):540–542. https://doi.org/10.1139/h2012-024
6. Firth J, Solmi M, Wootton RE, Vancampfort D, Schuch FB, Hoare E, Gilbody S, Torous J, Teasdale SB, Jackson SE, Smith L, Eaton M, Jacka FN, Veronese N, Marx W, Ashdown-Franks G, Siskind D, Sarris J, Rosenbaum S et al (2020) A meta-review of "lifestyle psychiatry": the role of exercise, smoking, diet and sleep in the prevention and treatment of mental disorders. World Psychiatry 19(3):360–380. https://doi.org/10.1002/wps.20773
7. Goff DC, Sullivan LM, McEvoy JP, Meyer JM, Nasrallah HA, Daumit GL, Lamberti S, D'Agostino RB, Stroup TS, Davis S, Lieberman JA (2005) A comparison of ten-year cardiac risk estimates in schizophrenia patients from the CATIE study and matched controls. Schizophr Res 80(1):45–53. https://doi.org/10.1016/j.schres.2005.08.010

8. Tam J, Warner KE, Meza R (2016) Smoking and the reduced life expectancy of individuals with serious mental illness. Am J Prev Med 51(6):958–966. https://doi.org/10.1016/j.amepre.2016.06.007
9. Goldfarb M, De Hert M, Detraux J, Di Palo K, Munir H, Music S, Piña I, Ringen PA (2022) Severe mental illness and cardiovascular disease. J Am Coll Cardiol 80(9):918–933. https://doi.org/10.1016/j.jacc.2022.06.017
10. Abou-Setta AM, Mousavi SS, Spooner C, Schouten JR, Pasichnyk D, Armijo-Olivo S, Beaith A, Seida JC, Dursun S, Newton AS, Hartling L (2012) First-generation versus second-generation antipsychotics in adults: comparative effectiveness. U.S Agency for Healthcare Research and Quality
11. Schuch FB, Vancampfort D (2021) Physical activity, exercise, and mental disorders: It is time to move on. Trends Psychiatry Psychother 43(3):177–184. https://doi.org/10.47626/2237-6089-2021-0237
12. Dauwan M, Begemann MJH, Heringa SM, Sommer IE (2016) Exercise Improves clinical symptoms, quality of life, global functioning, and depression in schizophrenia: a systematic review and meta-analysis. Schizophr Bull 42(3):588–599. https://doi.org/10.1093/schbul/sbv164
13. Dauwan M, Begemann MJH, Slot MIE, Lee EHM, Scheltens P, Sommer IEC (2021) Physical exercise improves quality of life, depressive symptoms, and cognition across chronic brain disorders: a transdiagnostic systematic review and meta-analysis of randomized controlled trials. J Neurol 268(4):1222–1246. https://doi.org/10.1007/s00415-019-09493-9
14. Rosenbaum S, Tiedemann A, Sherrington C, Curtis J, Ward PB (2014) Physical activity interventions for people with mental illness: a systematic review and meta-analysis. J Clin Psychiatry 75(9):14465. https://doi.org/10.4088/JCP.13r08765
15. Stubbs B, Vancampfort D, Hallgren M, Firth J, Veronese N, Solmi M, Brand S, Cordes J, Malchow B, Gerber M, Schmitt A, Correll CU, De Hert M, Gaughran F, Schneider F, Kinnafick F, Falkai P, Möller HJ, Kahl KG (2018) EPA guidance on physical activity as a treatment for severe mental illness: a meta-review of the evidence and position statement from the European Psychiatric Association (EPA), supported by the International Organization of Physical Therapists in Mental Health (IOPTMH). Eur Psychiatry J Assoc Eur Psychiatr 54:124–144. https://doi.org/10.1016/j.eurpsy.2018.07.004
16. Kandola A, Ashdown-Franks G, Hendrikse J, Sabiston CM, Stubbs B (2019) Physical activity and depression: towards understanding the antidepressant mechanisms of physical activity. Neurosci Biobehav Rev 107:525–539. https://doi.org/10.1016/j.neubiorev.2019.09.040
17. Lopez-Moral A, Munguia-Izquierdo D, Bueno-Antequera J (2024b) Exploring the role of physical exercise to improve cardiorespiratory fitness and muscular strength among individuals with severe mental disorder. In: Carneiro L, Schuch F (eds) Combining exercise and psychotherapy to treat mental health. IGI Global, pp. 182–198. https://doi.org/10.4018/978-1-6684-6040-5.ch008
18. Gurillo P, Jauhar S, Murray RM, MacCabe JH (2015) Does tobacco use cause psychosis? Systematic review and meta-analysis. Lancet Psychiatry 2(8):718–725. https://doi.org/10.1016/S2215-0366(15)00152-2
19. Gee B, Orchard F, Clarke E, Joy A, Clarke T, Reynolds S (2019) The effect of non-pharmacological sleep interventions on depression symptoms: a meta-analysis of randomised controlled trials. Sleep Med Rev 43:118–128. https://doi.org/10.1016/j.smrv.2018.09.004
20. Teasdale SB, Firth J (2020) Recommendations for dietetics in mental healthcare. J Hum Nutr Diet 33(2):149–150. https://doi.org/10.1111/jhn.12716
21. Vancampfort D, Rosenbaum S, Schuch FB, Ward PB, Probst M, Stubbs B (2016) Prevalence and predictors of treatment dropout from physical activity interventions in schizophrenia: a meta-analysis. Gen Hosp Psychiatry 39:15–23. https://doi.org/10.1016/j.genhosppsych.2015.11.008
22. Stubbs B, Vancampfort D, Rosenbaum S, Ward PB, Richards J, Soundy A, Veronese N, Solmi M, Schuch FB (2016) Dropout from exercise randomized controlled trials among people with

depression: a meta-analysis and meta regression. J Affect Disord 190:457–466. https://doi.org/10.1016/j.jad.2015.10.019
23. Vancampfort D, Sánchez CPR, Hallgren M, Schuch F, Firth J, Rosenbaum S, Van Damme T, Stubbs B (2021) Dropout from exercise randomized controlled trials among people with anxiety and stress-related disorders: a meta-analysis and meta-regression. J Affect Disord 282:996–1004. https://doi.org/10.1016/j.jad.2021.01.003
24. Sørensen M, Bentzen M, Farholm A (2020) Lessons learned from a physical activity intervention in psychiatric treatment: patient, staff, and leader perspectives. Front Psych 11:87. https://doi.org/10.3389/fpsyt.2020.00087
25. Curbelo Rodríguez R, Zarco Montejo P, Almodóvar González R, Flórez García M, Carmona Ortells L (2017) Barreras y facilitadores para la práctica del ejercicio físico en pacientes con espondiloartritis: Estudio cualitativo de grupos focales (EJES-3D). Reumatología Clínica 13(2):91–96. https://doi.org/10.1016/j.reuma.2016.03.004
26. Farina N, Williams A, Clarke K, Hughes LJ, Thomas S, Lowry RG, Banerjee S (2021) Barriers, motivators and facilitators of physical activity in people with dementia and their family carers in England: dyadic interviews. Aging Ment Health 25(6):1115–1124. https://doi.org/10.1080/13607863.2020.1727852
27. Wheeler AJ, Roennfeldt H, Slattery M, Krinks R, Stewart V (2018) Codesigned recommendations for increasing engagement in structured physical activity for people with serious mental health problems in Australia. Health Soc Care Community 26(6):860–870. https://doi.org/10.1111/hsc.12597
28. Firth J, Siddiqi N, Koyanagi AI, Siskind D, Rosenbaum S, Galletly C et al (2019) The Lancet Psychiatry Commission: a blueprint for protecting physical health in people with mental illness. Lancet Psychiatry 6(8):675–712
29. Firth J, Carney R, Pownall M, French P, Elliott R, Cotter J, Yung AR (2017) Challenges in implementing an exercise intervention within residential psychiatric care: A mixed methods study. Ment Health Phys Act 12:141–146. https://doi.org/10.1016/j.mhpa.2017.04.004
30. Korman NH, Shah S, Suetani S, Kendall K, Rosenbaum S, Dark F, Nadareishvili K, Siskind D (2018) Evaluating the feasibility of a pilot exercise intervention implemented within a residential rehabilitation unit for people with severe mental illness: GO HEART: (Group Occupational Health Exercise and Rehabilitation Treatment). Front. Psychiatry 9:343. https://doi.org/10.3389/fpsyt.2018.00343
31. Vancampfort D, Rosenbaum S, Probst M, Soundy A, Mitchell AJ, De Hert M et al (2015) Promotion of cardiorespiratory fitness in schizophrenia: a clinical overview and meta-analysis. Acta Psychiatr Scand 132:131–143. https://doi.org/10.1111/acps.12407
32. Vancampfort D, Rosenbaum S, Schuch F, Ward PB, Richards J, Mugisha J, Probst M, Stubbs, B (2017) Cardiorespiratory fitness in severe mental illness: a systematic review and meta-analysis. Sports Med (Auckland, N.Z.) 47(2):343–352. https://doi.org/10.1007/s40279-016-0574-1
33. Thomas P (2007) The stable patient with schizophrenia: from antipsychotic effectiveness to adherence. Eur Neuropsychopharmacol 17:S115–S122. https://doi.org/10.1016/j.euroneuro.2007.02.003
34. Binnie J, Boden Z (2016) Non-attendance at psychological therapy appointments. Mental Health Rev J 21:231–248. https://doi.org/10.1108/MHRJ-12-2015-0038
35. Lincoln T, Rief W, Westermann S, Ziegler M, Kesting M, Heibach E et al (2014) Who stays, who benefits? Predicting dropout and change in cognitive behaviour therapy for psychosis. Psychiatry Res 216:198–205. https://doi.org/10.1016/j.psychres.2014.02.012
36. Stanton R, Donohue T, Garnon M, Happell B (2016) Participation in and satisfaction with an exercise program for inpatient mental health consumers. Perspect Psychiatr Care 52(1):62–67. https://doi.org/10.1111/ppc.12108
37. Deenik J, Tenback D, Hendriksen I, Tak E, Van Harten P (2018) Multidisciplinary lifestyle-enhancing treatment for inpatients with severe mental illness (multi-study): effects on physical health, psychotic symptoms, quality of life and functioning. Schizophrenia Bull 44(suppl_1):S407–S407

38. Watkins A, Denney-Wilson E, Curtis J, Teasdale S, Rosenbaum S, Ward PB, Stein-Parbury J (2020) Keeping the body in mind: a qualitative analysis of the experiences of people experiencing first-episode psychosis participating in a lifestyle intervention programme. Int J Ment Health Nurs 29(2):278–289. https://doi.org/10.1111/inm.12683
39. Korman N, Fox H, Skinner T, Dodd C, Suetani S, Chapman J, Parker S, Dark F, Collins C, Rosenbaum S, Siskind D (2020) Feasibility and acceptability of a student-led lifestyle (diet and exercise) intervention within a residential rehabilitation setting for people with severe mental illness, GO HEART (Group Occupation, Health, Exercise and Rehabilitation Treatment). Front Psychiatry 11:319. https://doi.org/10.3389/fpsyt.2020.00319
40. Archie S, Wilson J, Osborne S, Hobbs H, McNiven J (2003) Pilot study: access to fitness facility and exercise levels in olanzapine-treated patients. Can J Psychiatry 48:628–632. https://doi.org/10.1177/070674370304800910
41. Firth J, Rosenbaum S, Stubbs B, Gorczynski P, Yung AR, Vancampfort D (2016) Motivating factors and barriers towards exercise in severe mental illness: a systematic review and meta-analysis. Psychol Med 46(14):2869–2881. https://doi.org/10.1017/S0033291716001732
42. Lopez-Mora A, Munguia-Izquierdo D, Bueno-Antequera J (2024) Barriers, facilitators, and experiences of physical exercise programs in outpatient mental health services: a global perspective of patients, staff, and trainers—the psychiactive project. In: Carneiro L, Schuch F (eds) Combining exercise and psychotherapy to treat mental health. IGI Global, pp. 140–160. https://doi.org/10.4018/978-1-6684-6040-5.ch006
43. Romain AJ, Longpré-Poirier C, Tannous M, Abdel-Baki A (2020) Physical activity for patients with severe mental illness: preferences, barriers and perceptions of counselling. Sci Sports 35(5):289–299. https://doi.org/10.1016/j.scispo.2020.03.005
44. Chapman JJ, Fraser SJ, Brown WJ, Burton NW (2016) Physical activity preferences, motivators, barriers and attitudes of adults with mental illness. J Ment Health 25(5):448–454. https://doi.org/10.3109/09638237.2016.1167847
45. Bernard P, Romain AJ, Esseul E, Artigusse M, Poy Y, Baghdadli A, Ninot G (2013) Barrières et motivation à l'activité physique chez l'adulte atteint de schizophrénie: revue de littérature systématique. Sci Sports 28(5):247–252. https://doi.org/10.1016/j.scispo.2013.02.005
46. Glover CM, Ferron JC, Whitley R (2013) Barriers to exercise among people with severe mental illnesses. Psychiatr Rehabil J 36(1):45–47
47. Kandola AA, Osborn DPJ (2022) Physical activity as an intervention in severe mental illness. BJPsych Adv 28(2):112–121. https://doi.org/10.1192/bja.2021.33
48. Soundy A, Stubbs B, Probst M, Hemmings L, Vancampfort D (2014) Barriers to and facilitators of physical activity among persons with schizophrenia: a survey of physical therapists. Psychiatr Serv 65(5):693–696. https://doi.org/10.1176/appi.ps.201300276
49. Rastad C, Martin C, Åsenlöf P (2014) Barriers, benefits, and strategies for physical activity in patients with schizophrenia. Phys Ther 94(10):1467–1479. https://doi.org/10.2522/ptj.20120443
50. Fibbins H, Lederman O, Rosenbaum S (2021) Physical activity and severe mental illness. In: Zenko Z, Jones L (eds) Essentials of exercise and sport psychology: an open access textbook. Society for Transparency, Openness, and Replication in Kinesiology, pp 385–408. https://doi.org/10.51224/B1017
51. Vancampfort D, De Hert M, Probst M, Firth J, Myin-Germeys I, Van Winkel R, Naisiga A, Basangwa D, Mugisha J (2017) Interest, competence, appearance, fitness and social relatedness as motives for physical activity in Ugandan outpatients with psychosis. Ment Health Phys Act 13:94–99. https://doi.org/10.1016/j.mhpa.2017.05.001
52. Méndez-Aguado C, Cangas AJ, Aguilar-Parra JM, Lirola MJ (2023) Benefits, facilitators and barrier reductions in physical activity programmes for people with severe mental disorder: a systematic review. Healthcare 11(9):1215. https://doi.org/10.3390/healthcare11091215
53. Vancampfort D, De Hert M, Broderick J, Lederman O, Firth J, Rosenbaum S, Probst M (2018) Is autonomous motivation the key to maintaining an active lifestyle in first-episode psychosis? Early Interv Psychiatry 12(5):821–827. https://doi.org/10.1111/eip.12373

54. Ryan RM, Deci EL (2020) Intrinsic and extrinsic motivation from a self-determination theory perspective: definitions, theory, practices, and future directions. Contemp Educ Psychol. https://doi.org/10.1016/j.cedpsych.2020.101860
55. Deci EL, Ryan RM (2000) The "what" and "why" of goal pursuits: human needs and the self-determination of behavior. Psychol Inq 11(4):227–268. https://doi.org/10.1207/S15327965PLI1104_01
56. Vancampfort D, Stubbs B, Venigalla SK, Probst M (2015) Adopting and maintaining physical activity behaviours in people with severe mental illness: the importance of autonomous motivation. Prev Med 81:216–220. https://doi.org/10.1016/j.ypmed.2015.09.006
57. Michie S, Carey RN, Johnston M, Rothman AJ, De Bruin M, Kelly MP, Connell LE (2018) From theory-inspired to theory-based interventions: a protocol for developing and testing a methodology for linking behaviour change techniques to theoretical mechanisms of action. Ann Behav Med 52(6):501–512
58. Carraca E, Encantado J, Battista F, Beaulieu K, Blundell J, Busetto L, Van Baak M, Dicker D, Ermolao A, Farpour-Lambert N, Pramono A, Woodward E, Bellicha A, Oppert JM (2021) Effective behavior change techniques to promote physical activity in adults with overweight or obesity: a systematic review and meta-analysis. Obes Rev 22(S4):13258. https://doi.org/10.1111/obr.13258
59. Duff OM, Walsh DM, Furlong BA, O'Connor NE, Moran KA, Woods CB (2017) Behavior change techniques in physical activity eHealth interventions for people with cardiovascular disease: systematic review. J Med Internet Res 19(8):e281. https://doi.org/10.2196/jmir.7782
60. French DP, Olander EK, Chisholm A, McSharry J (2014) Which behaviour change techniques are most effective at increasing older adults' self-efficacy and physical activity behaviour? A systematic review. Ann Behav Med 48(2):225–234. https://doi.org/10.1007/s12160-014-9593-z
61. Hoedjes M, Van Stralen MM, Joe STA, Rookus M, Van Leeuwen F, Michie S, Seidell JC, Kampman E (2017) Toward the optimal strategy for sustained weight loss in overweight cancer survivors: a systematic review of the literature. J Cancer Surviv 11(3):360–385. https://doi.org/10.1007/s11764-016-0594-8
62. Knittle K, Nurmi J, Crutzen R, Hankonen N, Beattie M, Dombrowski SU (2018) How can interventions increase motivation for physical activity? A systematic review and meta-analysis. Health Psychol Rev 12(3):211–230. https://doi.org/10.1080/17437199.2018.1435299
63. Michie S, Abraham C, Whittington C, McAteer J, Gupta S (2009) Effective techniques in healthy eating and physical activity interventions: a meta-regression. Health Psychol 28(6):690–701. https://doi.org/10.1037/a0016136
64. Thal SB, Maunz LA, Quested E, Bright SJ, Myers B, Ntoumanis N (2022) Behavior change techniques in physical activity interventions for adults with substance use disorders: a systematic review. Psychol Addict Behav. https://doi.org/10.1037/adb0000842
65. Cody R, Beck J, Brand S, Donath L, Faude O, Hatzinger M, Imboden C, Kreppke J-N, Lang UE, Ludyga S, Mans S, Mikoteit T, Oswald A, Schweinfurth N, Zahner L, Gerber M (2023) Short-term outcomes of physical activity counseling in in-patients with major depressive disorder: Results from the PACINPAT randomized controlled trial. Front Psych. https://doi.org/10.3389/fpsyt.2022.1045158
66. Suen YN, Lo LHL, Lee EHM, Hui CLM, Chan SKW, Chang WC, Chen EYH (2022) Motivational coaching augmentation of exercise intervention for early psychotic disorders: a randomised controlled trial. Aust N Z J Psychiatry 56(10):1277–1286. https://doi.org/10.1177/00048674211061496
67. Seymour J, Pratt G, Patterson S, Korman N, Rebar A, Tillston S, Chapman J (2022) Changes in self-determined motivation for exercise in people with mental illness participating in a community-based exercise service in Australia. Health Soc Care Community 30(5):e1611–e1624. https://doi.org/10.1111/hsc.13588
68. Zeibig JM, Seiffer B, Sudeck G, Rösel I, Hautzinger M, Wolf S (2021) Transdiagnostic efficacy of a group exercise intervention for outpatients with heterogenous psychiatric disorders:

a randomized controlled trial. BMC Psychiatry 21(1). https://doi.org/10.1186/s12888-021-03307-x
69. Lederman O, Ward PB, Rosenbaum S, Maloney C, Watkin A, Teasdale S, Morell R, Curtis J (2020) Stepping up early treatment for help-seeking youth with at-risk mental states: Feasibility and acceptability of a real-world exercise program. Early Interv Psychiatry 14(4):450–462. https://doi.org/10.1111/eip.12871
70. Ringen PA, Falk RS, Antonsen B, Faerden A, Mamen A, Rognli EB, Solberg DK, Martinsen EW, Andreassen OA (2018) Using motivational techniques to reduce cardiometabolic risk factors in long term psychiatric inpatients: a naturalistic interventional study. BMC Psychiatry 18(1):255. https://doi.org/10.1186/s12888-018-1832-6
71. Brooke LE, Gucciardi DF, Ntoumanis N, Chapman MT, Lines RLJ, Perry Y, Gilbey D, Formby T, Phillips T, Lin A (2022) Enhancing functional recovery for young people recovering from first episode psychosis via sport-based life skills training: outcomes of a feasibility and pilot study. Health Psychol Behav Med 10(1):1136–1158. https://doi.org/10.1080/21642850.2022.2147073
72. Orleans-Pobee M, Browne J, Ludwig K, Merritt C, Battaglini CL, Jarskog LF, Sheeran P, Penn DL (2022) Physical Activity Can Enhance Life (PACE-Life): results from a 10-week walking intervention for individuals with schizophrenia spectrum disorders. J Ment Health 31(3):357–365. https://doi.org/10.1080/09638237.2021.1875403
73. Jiménez-Zazo F, Hernández-Martínez A, Romero-Blanco C, Cabanillas E, Pérez-Gómez J, Ara I, Casajús JA, Vicente-Rodríguez G, Gesteiro E, González-Gross M, Aznar S (2022) Physical activity adherence related to body composition and physical fitness in spanish older adults: 8 years-longitudinal EXERNET study. Front Psychol. https://doi.org/10.3389/fpsyg.2022.858312
74. Chau RMW, Tsui AYY, Wong EYW, Cheung EYY, Chan DYC, Lau PMY, Ng RMK (2020) Effectiveness of a structured physical rehabilitation program on the physical fitness, mental health and pain for Chinese patients with major depressive disorders in Hong Kong: a randomized controlled trial with 9-month follow-up outcomes. Disabil Rehabil 44(8):1294–1304. https://doi.org/10.1080/09638288.2020.1800833
75. Imboden C, Gerber M, Beck J, Holsboer-Trachsler E, Pühse U, Hatzinger M (2020) Aerobic exercise or stretching as add-on to inpatient treatment of depression: Similar antidepressant effects on depressive symptoms and larger effects on working memory for aerobic exercise alone. J Affect Disord 276:866–876. https://doi.org/10.1016/j.jad.2020.07.052
76. Curcic D, Stojmenovic T, Djukic-Dejanovic S, Dikic N, Vesic-Vukasinovic M, Radivojevic N, Andjelkovic M, Borovcanin M, Djokic G (2017) Positive impact of prescribed physical activity on symptoms of schizophrenia: randomized clinical trial. Psychiatria Danubina 29(4):459–465. https://doi.org/10.24869/psyd.2017.459
77. Şenormancı G, Korkmaz N, Şenormancı Ö, Uğur S, Topsaç M, Gültekin O (2021) Effects of exercise on resilience, insight and functionality in patients with chronic schizophrenia in a psychiatric nursing home setting: a randomized controlled trial. Issues Ment Health Nurs 42(7):690–698. https://doi.org/10.1080/01612840.2020.1847221

Mental Health Disparities in the Latino Community: Focus on College Students

15

Frank Cárdenas

Abstract

Latinos are the largest minority group in the United States with a population of 63.7 million; 19.1% of the total U.S. population. By 2060, the number of Latino people in the United States is projected to grow to 119 million, or 28.6% of the population. They contribute to the economy, cultural diversity, and health of the nation. The U.S. Latino community is very diverse and includes people from many different nations and races. While many have lived in the U.S. for many generations, others are recent immigrants who may face inequities in socioeconomic status, education, and access to healthcare services. While Latinos are afflicted with mental illness at about the same rate as non-Latino Whites, 22% compared with 23.9%, only 36% of Latinos received mental health services compared to 52% of non-Latino Whites. Latinos are the youngest major racial/ethnic group in the U.S.; one-third of the nation's Latino population is younger than 18. Among all U.S. undergraduate and graduate students, the share of Latinos increased from 14% in 2010 to 20% in 2022, slightly higher than the Latino share of the total population. In addition to the typical college stressors, Latino students likely experience unique challenges as members of an ethnic minority group that may negatively affect their psychological well-being. Assessing Latino college students' characteristics in addition to their health status and health needs is key to informing health policy formulation and program implementation aimed to reduce and manage stress.

Keywords

Mental health • Latino students • Barriers • Treatment • Interdisciplinary approach

F. Cárdenas (✉)
Department of Management, California State University Los Angeles, Los Angeles, California, USA
e-mail: frank@frankcardenas.com

B. W. K. Son (ed.), *The Landscape of Global Health Inequity*, Integrated Science 22,
https://doi.org/10.1007/978-3-031-60502-4_15

1 Introduction

Latino college students are experiencing elevated levels of stress and mental health issues, significantly impacting their college experiences and jeopardizing their degree attainment, in fact, 42% of Latino students in 2023 said that they had considered dropping out of college in the prior six months, a higher rate than their White or Black peers, with emotional stress and personal mental health being their top reasons for considering leaving school, making it a more important factor than cost or coursework difficulty [1]. Various barriers to mental health services, particularly evident in Latino cultures, contribute to this issue. The stigma surrounding discussions about mental health issues and seeking treatment results in many Latino college students enduring their struggles in silence. Motivated by a decade of witnessing this troubling trend at a predominantly Latino campus, the paper delves deeper into the phenomena, aiming to understand its causes and explore potential strategies to alleviate the suffering. Recognizing the interdisciplinary nature of public health issues, the investigation extends into psychiatric, demographic, and other social sciences. This approach aligns with the ethos of this paper, which is dedicated to interdisciplinary analyses [2].

While the focus of this chapter deals with mental health issues and disparities specific to Latino college students, we should remain mindful that mental health problems are intensifying among the U.S. college student population as a whole. According to a report by the *Healthy Minds Study*, more than 60% of college students met the criteria for at least one mental health problem during the 2020–2021 school year, continuing a trend that has been on the rise for the past decade. It is also increasingly evident that the traditional counseling center model is ill-suited to address this burgeoning crisis [3].

2 Background on the Latino Community

The Latino population in the United States is very diverse and includes any person of Cuban, Mexican, Puerto Rican, South or Central American, or other Spanish culture or origin, regardless of race. More than 19.1% of the U.S. population (63.7 million) self-identify as Latino or Hispanic, up from 50.5 million in 2010, making people of Latino origin the nation's largest racial/ethnic minority. The 26% increase in the Latino population was faster than the nation's 8% growth rate [4]. The number of Latino people in the United States is projected to grow to 119 million, or 28.6% of the population, by 2060 [4]. Projections from the 2020 Census Bureau suggest that the 2060 life expectancies at birth for Latinos are 86.5 years, with 88.2 years for women, and 84.8 years for men. For non-Latino whites, the projected life expectancies are 85.6 years, with 87.4 years for women, and 84.0 years for men [6]. Latinos are also the youngest ethnic minority group in the country: 17.9 million Latinos, about a third of the Latino population, are under the age of 18. Nearly half of all U.S.-born Latinos are younger than 18 [7].

In 2020, among Latino subgroups, Mexicans ranked as the largest at 61.6%. Following this group were Puerto Ricans (9.6%), Central Americans (9.3%), South Americans (6.4%), other Latinos (including Spanish) (5.8%), and Cubans (3.9%). In 2020, states with the largest Latino populations were Arizona, California, Colorado, Florida, Illinois, Nevada, New Jersey, New Mexico, New York, and Texas [8]. According to the U.S. Bureau of Labor Statistics 2021 data, 22.9% of Latinos, in comparison to 15.0% non-Latino whites, worked within service occupations. About 24.5% of Latinos in comparison to 42.8% of whites worked in managerial or professional occupations. Among full-time year-round workers in 2020, the average Latino median household income was $55,321 in comparison to $74,912 for non-Latino white households. In 2020, the unemployment rate for Latinos was 10.4, as compared to 7.5 for non-Latino whites [9]. The poverty rate of Latinos is 76% higher (16.7%) than non-Latino White Americans (9.5%) [10].

Latino health is often shaped by factors such as language/cultural barriers, lack of access to preventive care, and the lack of health insurance [11]. According to 2019 data from the Centers for Disease Control and Prevention, some of the leading causes of death among Latinos include cancer, heart disease, unintentional injuries (accidents), stroke and other cerebrovascular diseases, diabetes, and Alzheimer's disease [12]. In 2020, the leading cause of death for Latinos was COVID-19. Some other causes of mortality that significantly affect Latinos include chronic lower respiratory diseases (including asthma and chronic obstructive pulmonary disease), liver disease, influenza and pneumonia, suicide, and kidney disease [13]. Latinos have higher rates of obesity than non-Latino whites [14]. There also are disparities among Latino subgroups. For instance, the rate of low birth weight infants is higher for the total Latino population in comparison to non-Latino whites, with Puerto Ricans having the highest rates of low birth weights among Latinos [15].

It is significant to note that Latinos have the highest uninsured rates of any racial or ethnic group within the United States. In 2020, the Census Bureau reported that 49.9% of Latinos had private insurance coverage, as compared to 73.9% for non-Latino whites. Among Latino subgroups, examples of coverage varied: 47.9% of Mexicans, 56.3% of Puerto Ricans, 57.4% of Cubans, and 41.7% of Central Americans. In 2020, 35.9% of all Latinos had Medicaid or public health insurance coverage, as compared to 33.8% for non-Latino whites. Public health insurance coverage varied among Latino subgroups: 36.4% of Mexicans, 43.7% of Puerto Ricans, 33.7 of Cubans, and 33.0% of Central Americans. Those without health insurance coverage varied among Latino subgroups: 20.3% of Mexicans, 8.0% of Puerto Ricans, 14.0% of Cubans, and 19.4% of Central Americans. In 2020, 18.3% of the Latino population was not covered by health insurance, as compared to 5.4% of the non-Latino white population per the Census Bureau report [12]. Low rates of insurance coverage for Latinos are likely to be a function of ethnicity, immigration status, and citizenship status.

3 Health Disparities

Health disparities or inequities are defined by the World Health Organization as "differences in health which are not only unnecessary and avoidable but are considered unfair and unjust" [16]. Significant health disparities in the United States manifest in a substantial disparity in life expectancy between counties with the highest and lowest life expectancies, amounting to 18.4 years for men and 14.3 years for women [17]. Numerous studies have demonstrated that race is a particularly strong and pervasive social determinant of health in the United States [18]. Latinos are disproportionately affected by poor conditions of daily life, shaped by structural and social position factors (such as macroeconomics, cultural values, income, education, occupation, and social support systems, including health services), known as social determinants of health (SDH). SDH is implicated in health inequalities, which are defined as health conditions that typically affect disproportionally certain socioeconomic, ethnic, and gender population subgroups [19]. SDH factors driving health inequalities, particularly in the Latino/Latino community include cultural background, employment, socioeconomic status, cultural background, and undocumented or foreign-born status [20]. Latinos had better health outcomes than whites for most analyzed health factors, despite facing worse socioeconomic barriers, but they had much higher death rates from diabetes, chronic liver disease/cirrhosis, and homicide, and a higher prevalence of obesity [21].

4 Behavioral and Mental Health of Latinos

4.1 Behavioral and Mental Health Issues

While Latinos are at lower risk of most psychiatric disorders compared with non-Latino whites, the incidence of self-reported mental illness among Latinos (nearly 22%) is very close to that reported by non-Latino Whites (23.9%) [22, 23]. Common mental health afflictions among Latino/Latinos are generalized anxiety disorder, major depression, post-traumatic stress disorder (PTSD), and excessive use of alcohol and drugs [24]. Latinos access mental health services at a much lower rate than non-Latino Whites; 36% of Latinos compared to 52% of non-Latino Whites [23]. Latinos are more likely to seek help for a mental health disorder from a primary care provider (10%) than a mental health specialist (5%). Poor communication with healthcare providers is often an issue, and there is a critical shortage of bilingual or Spanish-speaking mental health professionals [25].

Latino youth are more likely to have mental health issues than their peers. Twenty-two percent of Latino youth have depressive symptoms, a rate higher than any minority group besides Native American youth [26]. Latina youth have the highest rates of suicidal ideation and suicide attempts, and, while lower than Latinas, Latino adolescent males have higher rates of suicidal ideation and suicide attempts than White adolescents. Most suicide attempts by Latinos occurred before age 18, with Latinas twice as likely as males to attempt suicide [27]. Aggravating these stark figures is the fact that Latino children and adolescents are also less likely than their white peers to receive mental health treatment. While little data exists concerning suicides among Latino college students, in a psychosocial model of suicide risk among 156 Latino college students, Chang et al. [28] found significant evidence that the interaction of depression coupled with feelings of loneliness is an important predictor of suicide risk for this under-studied population.

4.2 Barriers and Facilitators to Mental Health Help-Seeking

A significant cultural barrier to mental health care access is the stigma associated with talking about mental illness as there is a perception, especially among older people, that the acknowledgment of problems may bring shame and embarrassment to the family [29]. Indeed, caregivers reported that they felt stigma and fear outweighed the benefits of receiving care for their Latino youth [30, 31]. Such stigmatization often leads to delayed treatment and prolonged suffering, as evidenced by studies suggesting that Latino parents might underreport their youth's mental health difficulties because of the stigma associated with identifying that their child has a mental health disorder [32].

In addition to the stigmatization of mental illness, Latinos experience barriers to behavioral health services and care such as experiences of racism and discrimination stemming from structural and systemic factors, language access issues, negative encounters with care providers, provider shortage due to the lack of providers from diverse racial/ethnic backgrounds, and lack of culturally competent resources and providers to meet cultural, social, and language-related needs [23].

Santana et al. [32] reviewed 27 peer-reviewed articles to understand the factors that Latino parents perceive to be barriers and facilitators to seeking help for their children's mental health symptoms and disorders. The result was a set of the barriers to help-seeking into seven themes: cultural factors and familial thoughts and beliefs; logistical; systemic and societal; problem recognition and identification; fears about seeking therapy; parental and familial behaviors and thoughts; and stigma and judgment. Additionally, seven themes related to facilitators emerged from the review: family, social, and community support; positive thinking and outlook; history of and familiarity with therapy; parental and child-related factors; demographic and logistical factors; school-related factors; and cultural factors [32].

4.3 The Reality of College Stress

Late adolescence and young adulthood are major developmental, transitional periods for youth in general, typically involving important physiological and psychological changes, including elevated stress [33–35]. This pattern is particularly true for college students as they experience more psychological stress compared to age cohorts who are not in college and compared to any other age cohort [36]. Three in four college students self-reported feeling stressed, while one in five college students reported stress-related suicidal ideation, according to the 2015 American College Health Association-National College Health Assessment report [37, 38].

A growing body of evidence suggests that academic-related stress plays a significant role in college. For instance, as many as 87% of college students surveyed across the United States cited education as their primary source of stress [37]. College students are exposed to academic stressors, such as an extensive academic course load, substantial studying, time management, classroom competition, financial concerns, familial pressures, and adapting to a new environment [38]. The consequences of experiencing college stress include lower academic achievement, college dropout, anxiety, loneliness, depression, hopelessness, and even suicidal ideation [39, 40].

4.4 College Stress Among Latinos

As more Latinos enter college, they similarly encounter the typical stressors of college life and young adulthood, but they often suffer additional stress uniquely associated with being an ethnic minority [41]. Over thirty years ago, Smedley et al. [42] coined the term "Minority stress" to refer to negative events experienced with being in a stigmatized group. *Minority stress* refers to negative experiences in the campus environment that students perceive to be linked to social, physical, or cultural attributes salient to the specific minority group individuals identify [42]. Smedley et al. [42] conceptualized minority stress among a minority group attending a predominantly White University as having five dimensions: social climate, difficulties in intergroup relations, discrimination, within-group pressures, and academic achievement stress. *University social climate stress* refers to perceptions of the campus environment as unwelcoming to members of the student's group. An example of *University social climate stress* is addressed by Castillo et al. [43] describing the stressful experience of acculturating into a college culture where the school is at least perceived to be modeled following inherently "White" values like individualism. *Intergroup stress* captures perceptions of negative relations among students from different racial and ethnic groups, primarily with White students. *Discrimination stress* refers to concerns related to personal experiences of prejudice and discrimination based on the person's group membership. *Within-group stress* captures perceived pressure to conform to the norms of the student's group regarding language, behaviors, and ways of thinking. *Achievement stress* reflects the student's concerns about the relative inadequacy of his or her academic preparation and ability that are recompounded by group membership and family social class background [43].

Arbona and Jimenez [41] reported that the stress of underrepresented groups was uniquely associated with depression. They also found that this stress compounded the stress related to academic achievement. Thus, while Latino college students experience general college stress, they are at risk of experiencing stress related to their racial/ethnic status which compounds their already existing college stress levels [41]. As such, Latino college students' characteristics should be considered in psychological interventions aimed at reducing and managing stress. Such an innovative and culturally sensitive approach is described by Terrazas-Carrillo and Garcia [44] in the establishment of stress management groups for Latino college students that provide psychoeducation, impart information, and teach stress-management skills within a Latino cultural framework. An emphasis is placed on the incorporation of content associated with acculturative stress and stress related to their status as a member of an underrepresented group to connect meaningfully with Latino college students [44].

5 Conclusion

The unique support needs of Latino college students warrant much more attention. Colleges need to continue to explore new approaches to mental health care and to provide support strategies for Latino students struggling with mental health issues that are informed by those stressors which are particular to the experience of being a Latino student. For example, administrations should carefully consider the realities of emotional stress and mental health [1], particularly depression and loneliness [28], in drop-out and suicide prevention as they design more culturally relevant and sensitive counseling programs. Leaders at colleges and universities must be involved in ensuring the integration of formal personnel training to support this unique and growing population of students, as the traditional counseling center model is not designed to address the mental health stressors and barriers to assistance faced by today's Latino collegians.

References

1. Gallup (2024) 2024 Lumina-Gallup State of Higher Education Study. https://news.gallup.com/poll/611093/black-hispanic-students-greatest-risk-leaving-program.aspx.
2. California State University Los Angeles (Fall 2023) Institutional Effectiveness, Enrollment. https://latabpubsrc.calstatela.edu/t/LAIR-PUBLIC/views/Enrollment/MultiracialMultiethnicCategories?%3Aembed=y&%3Aiid=1&%3AisGuestRedirectFromVizportal=y Accessed 1 Oct 2023
3. American Psychological Association (2022) Student mental health is in crisis. Campuses are rethinking their approach. https://www.apa.org/monitor/2022/10/mental-health-campus-care Accessed 7 Oct 2023
4. U.S. Census Bureau (2023) Facts for Features: Hispanic Heritage Month. U.S. Department of Commerce. https://www.census.gov/newsroom/facts-for-features/2023/hispanic-heritage-month.html#:~:text=63.7%20million,19.1%25%20of%20the%20total%20population.
5. U.S. Census Bureau (2015) Projections of the Size and Composition of the U.S. Population: 2014–2060. Department of Commerce. https://www.census.gov/content/dam/Census/library/publications/2015/demo/p25-1143.pdf

6. Medina LD, Sabo S, Vespa J (2020) Living Longer: Historical and Projected Life Expectancy in the United States, 1960 to 2060. Current Population Reports. Pp 25–1145. U.S. Census Bureau, Washington, DC. https://www.census.gov/content/dam/Census/library/publications/2020/demo/p25-1145.pdf
7. Patten E (2016) The Nation's Latino Population Is Defined by Its Youth. Pew Research Center's Hispanic Trends Project.
8. American Community Survey (n.d.) Five-Year Estimates Tables, B03001-Hispanic or Latino Origin by Specific Origin. https://data.census.gov/table/ACSDT5Y2020.B03001?y=2020&d=ACS%205-Year%20Estimates%20Detailed%20Tables
9. U.S. Bureau of Labor Statistics (November 2021) Labor force characteristics by race and ethnicity, 2020. Department of Labor. https://www.bls.gov/opub/reports/race-and-ethnicity/2020/home.htm
10. KFF (n.d.) State Health Facts, Poverty Rate by Race/Ethnicity. (Estimates based on the 2008–2022 American Community Survey, 1-Year Estimates). https://www.kff.org/other/state-indicator/poverty-rate-by-raceethnicity/?currentTimeframe=0&sortModel=%7B%22colId%22:%22Location%22,%22sort%22:%22asc%22%7D
11. Funk C, Lopez MH (2022) Hispanic Americans' Trust In And Engagement With Science, Hispanic Americans' Experiences with Health Care. Pew Research Center. https://www.pewresearch.org/science/2022/06/14/hispanic-americans-experiences-with-health-care/
12. Keisler-Starkey K, Bunch LN (Sept. 2021) U.S. Census Bureau Current Population Reports, P60–274, Health Insurance Coverage in the United States: 2020, U.S. Government Publishing Office, Washington, DC. https://www.census.gov/content/dam/Census/library/publications/2021/demo/p60-274.pdf
13. Centers for Disease Control and Prevention (n.d.) Underlying Causes of Death, 1999–2020. U.S. Department of Health & Human Services. https://wonder.cdc.gov/ucd-icd10.html
14. National Center for Health Statistics (2019) National Health and Nutrition Examination Survey. Table 26. Normal weight, overweight, and obesity among adults aged 20 and over, by selected characteristics: United States, selected years 1988–1994 through 2015–2018. https://www.cdc.gov/nchs/data/hus/2019/026-508.pdf
15. Osterman MJK. et al. (2022) Births: final data for 2020. National Vital Statistics Reports 70(17). https://doi.org/10.15620/cdc:112078
16. Carter-Pokras O, Baquet C (2016) What is a "health disparity"?. Public health reports. https://journals.sagepub.com/doi/abs/https://doi.org/10.1093/phr/117.5.426
17. Perez-Escamilla R (2013) Health disparities. In: Caballero B (ed) Encyclopedia of human nutrition, 3rd ed. Academic Press, pp 417–423. ISBN 9780123848857. https://doi.org/10.1016/B978-0-12-375083-9.00142-2. https://www.sciencedirect.com/science/article/pii/B9780123750839001422
18. Brandt EJ (2023) Social determinants of racial health inequities. The Lancet Public Health 8(6):e396–e397. https://www.thelancet.com/journals/lanpub/article/PIIS2468-2667(23)00100-7/fulltext#articleInformation
19. Howard G, Peace F, Howard VJ (2014) The contributions of selected diseases to disparities in death rates and years of life lost for racial/ethnic minorities in the United States, 1999–2010. Preventing Chronic Disease 11. https://www.cdc.gov/pcd/issues/2014/14_0138.htm
20. Nelson A (2002) Unequal treatment: confronting racial and ethnic disparities in health care. J National Med Assoc 94(8):666. https://www.ncbi.nlm.nih.gov/pmc/articles/PMC2594273/
21. Dominguez K, Penman-Aguilar A, Chang M, Moonesinghe R, Castellanos T, Rodriguez-Lainz A, Schieber A (2015) Vital signs: leading causes of death, prevalence of diseases and risk factors, and use of health services among Hispanics in the United States—2009–2013, Morbidity and mortality weekly report. U.S. Department of Health and Human Services. https://www.cdc.gov/mmwr/preview/mmwrhtml/mm6417a5.htm
22. Ward BW, Clarke TC, Nugent CN, Schiller JS (2016) Early release of selected estimates based on data from the 2015 national health interview survey. National Center for Health Statistics. http://www.cdc.gov/nchs/nhis.htm.

23. Substance Abuse and Mental Health Services Administration (2022) Key substance use and mental health indicators in the United States: results from the 2021 national survey on drug use and health (HHS Publication No. PEP22–07–01–005, NSDUH Series H-57). Center for Behavioral Health Statistics and Quality, Substance Abuse and Mental Health Services Administration. https://www.samhsa.gov/data/report/2021-nsduh-annual-national-report
24. National Alliance on Mental Illness [n.d.] Mental health challenges in Latino communities. https://namica.org/mental-health-challenges-in-latino-communities/
25. American Psychiatric Association (2017) Mental health disparities: hispanics and Latinos. https://www.psychiatry.org/File%20Library/Psychiatrists/Cultural-Competency/Mental-Health-Disparities/Mental-Health-Facts-for-Hispanic-Latino.pdf
26. Guzman A, Koons A, Postolache T (2009) Suicidal behavior in Latinos: focus on the youth. Int J Adolesc Med Health 21:431–439
27. Goldston DB et al (2008) Cultural considerations in adolescent suicide prevention and psychosocial treatment. Am Psychol 63:14–31
28. Chang EC, Chang OD, Lucas AG, Li M, Beavan CB, Eisner RS, McManamon BM, Rodriguez NS, Katamanin OM, Bourke EC, de la Fuente A, Cardeñoso O, Wu K, Yu EA, Jeglic EL, Hirsch JK (2019) Depression, loneliness, and suicide risk among Latino college students: a test of a psychosocial interaction model. Social Work 64(1):51–60. https://doi.org.mimas.calstatela.edu/10.1093/sw/swy052
29. Jimenez DE, Bartels SJ, Cardenas V, Alegría M (2013) Stigmatizing attitudes toward mental illness among racial/ethnic older adults in primary care. Int J Geriatric Psychiatry 28(10):1061–1068. https://doi.org/10.1002/gps.3928. Accessed from https://www.ncbi.nlm.nih.gov/pmc/articles/PMC3672370/
30. De Silva LED, Ponting C, Ramos G, Guevara MVC, Chavira DA (2020) Urban Latinx parents' attitudes towards mental health: mental health literacy and service use. Children and Youth Services Rev 109. https://doi.org/10.1016/j.childyouth.2019.104719
31. Roberts RE, Alegría M, Roberts CR, Chen IG (2005) Mental health problems of adolescents as reported by their caregivers: a comparison of European, African, and Latino Americans. J Behav Health Serv Res 32(1):1–13. https://doi.org/10.1007/BF02287324
32. Santana A, Williams C, Winter M, Sullivan T, Elias M, Moreno O (2023) A scoping review of barriers and facilitators to Latinx Caregivers' help-seeking for their children's mental health symptoms and disorders. J Child Fam Stud 32:1–18. https://doi.org/10.1007/s10826-023-02715-9
33. Matud MP, Díaz A, Bethencourt JM, Ibáñez I (2020) Stress and psychological distress in emerging adulthood: a gender analysis. J Clin Med 9:2859. https://doi.org/10.3390/jcm9092859
34. Scales PC, Benson PL, Oesterle S, Hill KG, Hawkins JD, Pashak TJ (2015) The dimensions of successful young adult development: a conceptual and measurement framework. Appl Dev Sci 20:150–174. https://doi.org/10.1080/10888691.2015.1082429
35. Spear LP (2000) The adolescent brain and age-related behavioral manifestations. Neurosci Biobehav Rev 24:417–463. https://doi.org/10.1016/S0149-7634(00)00014-2
36. Blanco C, Okuda M, Wright C, Hasin DS, Grant BF, Liu SM et al (2008) Mental health of college students and their non-college-attending peers: results from the national epidemiologic study on alcohol and related conditions. Arch Gen Psychiatry 65:1429–1437. https://doi.org/10.1001/archpsyc.65.12.1429
37. American Psychological Association (2020) Stress in AmericaTM2020: a national mental health crisis. Washington, DC, American Psychological Association
38. Liu CH, Stevens C, Wong S, Yasui M, Chen JA (2019) The prevalence and predictors of mental health diagnoses and suicide among U.S. college students: implications for addressing disparities in service use. Depress Anxiety 36:8–17. https://doi.org/10.1002/da.22830
39. Pascoe MC, Hetrick SE, Parker AG (2020) The impact of stress on students in secondary school and higher education. Int J Adolesc Youth 25:104–112. https://doi.org/10.1080/02673843.2019.1596823

40. Eisenberg D, Gollust SE, Golberstein E, Hefner JL (2007) Prevalence and correlates of depression, anxiety, and suicidality among university students. Am J Orthopsychiatry 77:534–542. https://doi.org/10.1037/0002-9432.77.4.534
41. Arbona C, Jimenez C (2013) Minority stress, ethnic identity, and depression among Latino college students. J Couns Psychol 61(1):162–168. https://doi.org/10.1037/a0034914
42. Smedley BD, Myers HF, Harell SP (1993) Minority-status stresses and college adjustment ethnic minority freshmen. J Higher Educ 64:434–452. https://doi.org/10.2307/2960051
43. Castillo LG, Waler JE, Zamboanga BL, Weisskirch RS, Park IJK, Navarro RL, Caraway SJ (2015) Gender matters: the influence of acculturation and acculturative stress on Latino college student depressive symptomology. J Latinx Psychol 3(1):40–55. https://doi.org/10.1037/lat0000030
44. Terrazas-Carrillo E, Garcia E (2022) Initial outcomes for a stress management psychoeducational group for Latino college students. J College Student Psychotherapy 1–12. https://doi.org.mimas.calstatela.edu/10.1080/87568225.2022.2133044

Inequity Versus Inequality in the One Health: Are We Doing Justice and to Whom?

16

Sandul Yasobant, Mayank Yadav, and Deepak Saxena

Abstract

One Health is a comprehensive approach that acknowledges the interconnection between human, animal, and environmental health. It emphasizes that the well-being of each element is reliant on the well-being of the others. This chapter explores the One Health approach, delving into its historical roots, ethical considerations, and the importance of equity and justice. It also highlights the significance of incorporating cultural perspectives into One Health initiatives. Ultimately, the chapter underscores the potential of One Health to create a more balanced, sustainable, and just world for all living beings.

Keywords

One health • Health disparities • Equity • Inequality • Ethics • Culture • Sustainability • Health justice

S. Yasobant (✉) · D. Saxena
Department of Public Health Sciences, Indian Institute of Public Health Gandhinagar (IIPHG), Gandhinagar 382042, India
e-mail: yasobant@iiphg.org

S. Yasobant · M. Yadav · D. Saxena
Center for One Health Education, Research, and Development (COHERD), Indian Institute of Public Health Gandhinagar (IIPHG), Gandhinagar 382042, India

S. Yasobant · D. Saxena
School of Epidemiology and Public Health, Datta Meghe Institute of Medical Sciences (DMIMS), Wardha 442004, India

S. Yasobant
Global Health, Institute for Hygiene and Public Health (IHPH), University Hospital Bonn, 53127 Bonn, Germany

B. W. K. Son (ed.), *The Landscape of Global Health Inequity*, Integrated Science 22,
https://doi.org/10.1007/978-3-031-60502-4_16

1 Introduction: Basic of One Health

One Health is an integrated, unifying approach that aims to sustainably balance and optimize the health of people, animals, and ecosystems. It recognizes the health of humans, domestic and wild animals, plants, and the wider environment (including ecosystems) are closely linked and interdependent. The approach mobilizes multiple sectors, disciplines, and communities at varying levels of society to work together to foster well-being and tackle threats to health and ecosystems while addressing the collective need for clean water, energy, and air, safe and nutritious food, taking action on climate change, and contributing to sustainable development [1].

One Health is founded on several key principles that include impartiality between sectors and disciplines, socio-political and multicultural parity (the belief that all people are equal and deserve equal rights and opportunities), inclusion and engagement of communities and marginalized voices, a socioecological equilibrium that seeks a harmonious balance between human-animal-environment interaction, acknowledging the importance of biodiversity, access to natural space and resources, and the intrinsic value of all living things within the ecosystem.

Additionally, One Health includes stewardship and the responsibility of humans to change their behavior and adopt sustainable solutions that recognize the importance of animal welfare and the integrity of the whole ecosystem, thus securing the well-being of current and future generations. Finally, transdisciplinary and multisectoral collaboration is another essential principle of One Health, which includes all relevant disciplines, both modern and traditional forms of knowledge, and a broad representative array of perspectives [1].

Now, let's look at the history of One Health, One Health is a relatively new term, but the idea dates back to ancient times. The notion that environmental factors can affect human health can be traced back to the Greek physician Hippocrates (c. 460 BCE–c. 370 BCE) in his book "On Airs, Waters, and Places." He promoted the idea that maintaining a clean environment was crucial for public health [2].

The Italian physician Giovanni Maria Lancisi (1654–1720) pioneered epidemiology, medicine, and veterinary science. He was fascinated with the role the physical environment played in the spread of disease in humans and animals. The concept that human, animal, and environmental health are interconnected was further developed during the French Revolution by Louis-René Villerme (1782–1863) and Alexandre Parent-Duchâtelet (1790–1835), who contributed to the development of the public hygiene specialty [3].

In the late nineteenth century, German physician and pathologist Rudolf Virchow (1821–1902) coined the term "zoonosis" and believed that there should be no dividing lines between animal and human medicine [4]. In the US in 1947, veterinarian James H. Steele advanced the concept by establishing the field of veterinary public health at the CDC. The phrase "One Medicine" was coined and promoted by Calvin W. Schwabe (1927–2006), a veterinary epidemiologist and parasitologist, in his book "Veterinary Medicine and Human Health" [5].

In 1996, Gary M. Tabor and his associates launched the Conservation Medicine: Ecological Health in Practice effort (Consortium for Conservation Medicine) with the Cummings School of Veterinary Medicine at Tufts University, Harvard Medical School's Center for Health and the Global Environment, and EcoHealth Alliance (formerly Wildlife Trust), with a focus on linking human, animal, and ecological health [6].

The term "One Health" was first mentioned in a story about Ebola hemorrhagic fever on April 7, 2003, when Rick Weiss of the Washington Post quoted William Karesh as saying, "Human or livestock or wildlife health can't be discussed in isolation anymore. There is just one health. And the solutions require everyone working together on all the different levels" [7]. As Schwabe (1984) stated, "There is no difference of paradigm between human and veterinary medicine, and is an extension of notions of comparative medicine. Both sciences share, as a general medicine, a common body of knowledge in anatomy, physiology, pathology, and the origin of diseases in all species" [8].

Over the past few decades, there have been more outbreaks of emerging infectious diseases (EIDs). Current illnesses that significantly affect society, economy, and health include severe acute respiratory syndrome (SARS), avian influenza, Ebola, and Coronavirus disease 2019 (COVID-19). More than 6.3 million people died as a result of COVID-19, which caused an unprecedented global health, economic, and social emergency and cost the world's economy roughly $13.8 trillion. Many of the world's infectious diseases have their origins in animals, particularly wildlife. The rise of emerging and re-emerging zoonotic diseases can be attributed to various factors, such as the encroachment of natural habitats, which disturbs the ecological balance and causes microbes to spill over from animals to humans. Therefore, it is essential to adopt an interdisciplinary approach that focuses on the human-animal-environment interface. The main lesson to be learned from past disease management is that it requires a joint effort from various stakeholders, following the One Health approach.

Various crucial aspects must be taken into consideration in One Health. These include ethical considerations in One Health research, equity among human, animal, and environmental health, and addressing any disparities that may exist within the One Health sphere. These are all significant concerns that raise questions about whether we are being fair, and if so, to whom.

2 Ethics in One Health Approach

The One Health perspective considers the well-being of humans, animals, and the environment. Humans seek to maximize their well-being, but we should also consider the rights and well-being of animals. Unfortunately, despite protective attitudes in most cultures and religions, animals are often reared, transported, and slaughtered in inhumane conditions, calling for stronger engagement in animal protection and welfare.

Animal biodiversity is important for stable ecosystem services, and livestock rearing helps maintain carbon sequestration in semi-arid areas. Animal diseases can threaten human health and food security through the transmission of zoonotic diseases or the loss of animal capacity for ploughing. Large parts of the world rely on moderate livestock use for inhabitation. Therefore, we must recognize the close linkage and interdependencies between human and animal health, and the maintenance of stable ecosystem services. Some of these services are seriously threatened by livestock-rearing methods and exploitative human activities [9].

Peter Rabinowitz, an occupational physician at Yale University, suggests that people should shift their perspective towards animals from "us versus them" to a more collaborative "shared risk" attitude between humans and animals. For instance, the high incidence of cancer among Beluga whales in the Saguenay Fjord in Canada is caused by their continuous exposure to wastewater from industries, making it an indicator of environmental quality. Hence, humans have a vested interest in preserving the environment's quality to safeguard both human and animal health [10].

From an integrative One Health, conservation biology, and ecosystem perspective, animals should be highly regarded and treated as a part of the overall effort to sustain ecosystem integrity and comprehensive well-being. This includes animal husbandry, animal traction, animal transport, slaughter practices, and wildlife conservation [11].

Livestock owners should receive regular training on the best animal welfare practices in their rearing system. Slaughtering practices should aim to minimize stress during animal handling. Meat consumption has grown significantly in the last few decades, and livestock plays a crucial role, particularly in the livelihood of millions of small-scale farmers, including their use for ploughing and transport. Although there is increasing research on livestock, companion animals, and wildlife in developing countries, there is almost no legislation regarding animal testing. Care should be taken to ensure that animal testing does not become an "offshore enterprise," allowing industrialized countries to evade stringent regulations [12].

Below are some examples that highlight the Ethical Aspects of One Health Approach, the article entitled Stamping Out Animal Culling: From Anthropocentrism to One Health Ethics mentioned that animal culling is a public health policy used to control animal populations for safety, conservation, or disease control. Hundreds of thousands of animals are killed daily worldwide to control outbreaks of zoonotic diseases. They can be wild, farm, or urban, and killed to protect humans or other animals [13].

Herten J et al. (2020) highlighted that many countries have early warning systems to control zoonotic diseases, however, these systems are ill-equipped to handle moral dilemmas such as culling healthy animals. Research in the Netherlands shows that while professionals involved in zoonotic disease control policies generally adhere to the One Health concept, an anthropocentric approach is dominant in practice, with public health being the primary moral value. This creates tension with One Health's thinking. Without ethical expertise in control systems for zoonotic diseases, there can be misconceptions of ethical principles, such as the precautionary principle, frequently used to justify disease control measures [14].

Anthony R et al. (2020) highlighted in their article that the One Health framework helps address human and animal health concerns. It can be used to promote animal welfare during disasters, but ethical considerations make implementation challenging. This discussion focuses on the framework's significance for animal disaster management, an ethics of care approach to human-animal conflicts, and insights from animal welfare science [15].

In the article entitled Supporting One Health for Pandemic Prevention: The Need for Ethical Innovation by Diller ER et al. (2023) narrated that Bioethics requires innovation to prevent zoonotic diseases. Pandemics like COVID-19 require balancing individual choice and societal responsibility. One Health promotes synergy between human, animal, and environmental health, but there's a power imbalance. A renewed concept of justice and flexible application of OH can sustain human and environmental health. Bioethics must play a key role in supporting this development [16].

The work by Nieuwland J et al. (2019) indicates that One Health is a framework that helps health professionals navigate the relationships between humans, animals, and ecosystems. It requires value judgments, making it important for students to engage critically from an ethical perspective. Veterinary ethics education can support this. One Health is a comprehensive concept that can be easily misused. Therefore, veterinary ethics education should encourage critical examination. One Health enriches veterinary ethics education by promoting collaboration across different disciplines and multicultural exchange of perspectives. Finally, One Health inspires veterinary ethics education to become more pedagogical by emphasizing interdependence [17].

Entitled Does One Health Require a Novel Ethical Framework by Johnson J and Degeling C. (2019) highlighted those emerging infectious diseases are a significant threat globally. The One Health (OH) approach is gaining scientific support to combat this problem. OH, recognizes the interconnections between human, animal, and environmental health. There is a growing interest in the philosophical and ethical dimensions of OH. While a new ethical framework is not necessarily required, creating one that aligns with OH could provide benefits. OH, advocates can still adhere to existing ethical frameworks [18].

Beever J and Morar N. (2019) also mentioned in the article entitled The Epistemic and Ethical Onus of 'One Health', that the One Health paradigm's practical reach and ethical impact depend on distinguishing between interconnected and interdependent factors among human, non-human, and environmental health. While interconnection does not necessarily mean interdependence, the existing One Health literature does not differentiate between these concepts sufficiently. As a result, stakeholders either ignore such distinctions or embrace bioethically untenable positions. Therefore, careful conceptual differentiation is crucial to prevent One Health stakeholders from overreaching or underreaching the practical and ethical boundaries of this developing paradigm [19].

Coghlan S et al. (2021) highlighted in their article that a bolder One Health: expanding the Moral Circle to Optimize Health for All. The One Health philosophy is a groundbreaking approach to improving health that challenges centuries-old assumptions about well-being. It is widely regarded as the best solution for mitigating human health problems, including pandemic zoonotic diseases. If we don't make big changes to the status quo, there will be great suffering and ill-health. However, One Health is not radical enough, even in its more ambitious forms. It has not embraced the emerging philosophical view that historical anthropocentrism is an unfounded ethical prejudice against other animals. This paper argues that One Health should be more imaginative and adventurous in its core philosophy and ultimately in its recommendations and activities. It must expand the circle of moral concern beyond a narrow focus on human interests to include nonhuman beings and the environment. On this bolder agenda, progressive ethical and practical thinking converge to benefit the planet and its diverse inhabitants, both human and nonhuman [20].

3 Equity Aspects of Human-Animal-Environment Health

In the One Health context, it is essential to consider equity to address the interrelated health of humans, animals, and the environment. Equity means ensuring a fair and impartial distribution of resources, opportunities, and burdens among different populations. To achieve equity in One Health initiatives, it is necessary to recognize and address disparities in access to healthcare, environmental resources, and animal welfare [21]. Vulnerable communities, including low-income populations and marginalized groups, are often more affected by zoonotic diseases, environmental degradation, and lack of access to veterinary care. Promoting equity requires implementing inclusive policies and interventions that prioritize the needs of these communities, empower local stakeholders, and address underlying social determinants of health. By promoting equity, One Health approaches can more effectively protect the health of all living beings and promote sustainable relationships between humans, animals, and the environment [22].

Some Articles also discuss the equity aspects of human-animal-environment health. In their article entitled One Health Contributions Towards More Effective and Equitable Approaches to Health in Low- and Middle-income Countries by Cleaveland S et al. (2017) mentioned that Preventive 'One Health' interventions, e.g. interventions in animal populations that generate human health benefits, may provide a useful approach to overcoming some of these challenges. Effective strategies, such as animal vaccination, already exist for the prevention, control, and elimination of many endemic zoonoses, including rabies, and several livestock zoonoses (e.g. brucellosis, leptospirosis, Q fever) that are important causes of human febrile illness and livestock productivity losses in low- and middle-income countries. We make the case that, for these diseases, One Health interventions have the potential to be more effective and generate more equitable benefits for human health and livelihoods, particularly in rural areas, than approaches that rely exclusively on the treatment of human cases [23].

An article entitled One Health Action for Health Security and Equity also focuses on the importance of integrated One Health approaches to reduce the likelihood and impacts of emerging infectious disease outbreaks has been widely accepted, as the *Lancet* series on One Health emphasizes. However, implementing One Health requires transdisciplinary approaches, with a systemic focus on the health of animals, humans, and ecosystems worldwide, and potential solutions that are equitable, inclusive, and sustainable [24].

4 Inequality Within One Health Domain

Health disparities are preventable diseases that arise due to social and systemic issues. Historians will likely focus on health disparities related to socio-economic status, access to healthcare, education level, environment, politics, and discrimination when writing about the pandemic in the future.

A novel framework called 'One Health Disparities' integrates perspectives on health disparities with One Health. The environment is a vital component of One Health and social determinants of health, where the social environment is essential for understanding many health disparities. Expanding the One Health framework to include aspects of the social environment can help explain variation in health disparities for zoonotic diseases [25].

To understand how One Health Disparities informs our understanding of the coronavirus pandemic, we apply it to three critical components of infectious disease transmission and epidemiology: exposure, susceptibility, and disease expression. For transmission of any infectious disease to occur, an individual must be exposed to the infectious agent and susceptible to it. Once infected, individuals vary in their expression of the disease. One Health Disparities aims to understand how disparities emerge at each of these stages by incorporating perspectives from social determinants of health, One Health, and evolutionary medicine [26].

Some more Articles also discuss the inequality within One Health Domains.

Eliakimu ES and Mans L. (2022) highlighted in their article entitled Addressing Inequalities Toward Inclusive Governance for Achieving One Health: A Rapid Review that inequalities have a significant impact on global health and well-being. An inclusive approach to governance is necessary to tackle these inequalities. However, COVID-19 has worsened the situation, making it difficult to address inequalities through One Health. While we have identified key strategies, they require strong policies at global, regional, and country levels, informed by policy sciences. Strengthening research capacity is also crucial. Therefore, inclusive governance is essential to achieve One Health [27]. Solis A and Nunn CL (2021) in their article One Health Disparities and COVID-19 also discuss that One Health focuses on interconnections between human, animal, and environmental health. We propose that social environments are also important to One Health and help illuminate disparities in the coronavirus pandemic, including its origins, transmission, and susceptibility among humans, and spillback to other species. We call this framework One Health Disparities [28].

5 Health Justice Within One Health

Lysaght T, Capps B, Bailey M, et al. (2017) mentioned that Integrating health justice into the One Health approach is essential for achieving overall well-being across human, animal, and environmental spheres. Although the One Health approach recognizes the interconnectedness of these realms, it often fails to consider the social and structural inequities that disproportionately affect certain communities [29]. Health justice frameworks can help us understand how existing power structures and resource distribution patterns shape health outcomes for different groups. This means acknowledging how factors such as poverty, racism, and environmental degradation create vulnerabilities to zoonotic diseases, food insecurity, and mental health issues. By integrating principles of environmental justice and community engagement into One Health initiatives, we can ensure equitable access to healthcare, sanitation, and healthy environments for everyone. Ultimately, this will lead to a more sustainable and fair future [30].

Some more Articles also discuss health justice within One Health. The article entitled Forming International Instrument Through One Health Approach for Health Justice highlighted that COVID-19 has proven that the health of humans is interconnected with the health of animals and the environment. International instruments have been formed to address this issue and reach a global understanding of how to deal with pandemics. This approach is known as the One Health approach for health justice. Disparities in COVID-19 infections, disease outcomes, and access to healthcare were found to be linked not only to healthcare system discrimination and inequality but also to inequity throughout society. The purpose of this research is to discuss the importance of One Health for health justice as a concept and approach to forming international instruments. The One Health approach recognizes that human health and the environment are deeply interconnected. This approach advocates for interdisciplinary and cross-sectoral cooperation, such as health diplomacy, as a forum for negotiation between various states [30].

In the article entitled Punishment to Support: The Need to Align Animal Control Enforcement with the Human Social Justice Movement by Hawes SM, Hupe T, Morris KN discuss about the social justice movement is striving to raise awareness that addressing public health and safety concerns requires addressing social justice issues that limit communities' capabilities. One Health and One Welfare frameworks can provide a more comprehensive approach to ensuring public health and safety that prevents further harm to underserved and socially disadvantaged populations, particularly low-income communities and people of color. By moving animal control policies from punishment to support, we can recognize the physical and emotional benefits of the human-animal bond and incorporate animal control agencies into a more robust system that supports pet ownership [31].

Murray MH, et al. (2022) mentioned in their article One Health for All: Advancing Human and Ecosystem Health in Cities by Integrating an Environmental Justice Lens that we are currently facing several global threats to public health and ecosystem function that reveal the close links between human and wildlife health. These challenges are particularly urgent in cities, where social-ecological interactions are more pronounced. The One Health concept helps to promote the health and well-being of urban communities and ecosystems by providing an organizing framework. However, for One Health to be truly effective, it must take into account societal inequities in environmental dis-amenities, exposures, and policies. These inequities affect all One Health interfaces, including the distribution of ecosystem services and disservices, the frequency and nature of human–wildlife interactions, and the effects of land use over time. A One Health perspective can help identify areas where an environmental justice lens should be incorporated, and we can conclude with recommendations for future work. Given the increasing social, political, and environmental unrest, we urgently need One Health solutions that are informed by environmental justice principles to help build healthier, more resilient cities [32].

6 Indian Culture and Sustainable Practice in One Health

Indigenous communities, influenced by their cultural traditions, are generally more mindful of their environment and more committed to preserving its ecological balance. This is both a practical matter for safeguarding their livelihoods and a matter of reverence for the sanctity of special places. In coastal Maharashtra, there are sacred groves (devrai) that are highly valued due to such traditions. These groves may have a temple, shrine, monastery, or burial ground, and they have served as havens for wildlife and biodiversity. Within the grove, it is strictly prohibited to cut plants or even the living branch of a tree, and the local population often feels accountable for their protection [33]. Nair S. et al. (2021) mentioned that Social, cultural, and religious studies recognize the importance of various other places, particularly those where rivers and their waters have the power to heal and restore. Commentaries on the literature of the Mahānubhāv religious sect, founded by Chakradhar in the thirteenth century, extol the psychological and health benefits of being in Maharashtra. It is believed to not only promote religious pursuits but also to be good for one's well-being and moral character [34].

Weiss MG, Agashe M, and Gupte M (2021) also highlighted that the Ayurvedic medical tradition is rooted in cultural ecological values and environmental harmony. The Charaka Saṃhitā explains the relationship between the individual and the world, as well as the ultimate emancipation from the world. The fieldwork and study of varied landscapes, non-herbal medicines, zoology, and pharmacy show how the setting is related to health and remedies. The natural world provides remedies for medical problems and overall health [35].

Ayurveda, a traditional Indian system of medicine, has dedicated sections that address the health problems resulting from interactions with animals. The texts discuss the effects of poisons and venoms from snakes, scorpions, and spiders. Exposure to the teeth, excreta, or semen of various mammals can also be toxic. The texts elaborate on the effects of poison in the semen of rats and mice and their treatment. In sections describing animal bites, two of the three major Ayurvedic texts—the Suśruta Saṃhitā (Chap. 7 of the Kalpasthāna section) and the Aṣṭāṅgahṛdaya Saṃhitā (chapter 38 of the Uttarasthāna section)—include accounts of rabies. The texts refer to rabies as hydrophobia (jaltrāsa), a term based on well-known symptoms, including fear of water. The condition was attributed to the bites of dogs and jackals, as well as other species, including hyenas, bears, and tigers. Affected persons may imitate the behaviors of the biting animal, and the condition is fatal for them [35].

Indian culture is known for its unique practices that have been passed down through generations. One such practice is the use of bamboo leaves as plates for meals, which not only adds a natural touch but also has several health benefits. Another traditional practice is the use of mud pots to store drinking water, as it keeps the water naturally cool and enhances its taste. Caring for animals is an integral part of the Indian way of life, especially for farmers who rely on them for their livelihood. Cows, in particular, are considered sacred and are worshipped in India. These cultural practices reflect the deep-rooted beliefs and values that Indians have cherished for centuries [35].

7 Conclusion

The One Health approach is a comprehensive and holistic strategy that emphasizes the interconnectedness of human health, animal health, and the environment. This approach recognizes that the health of humans, animals, and the environment is interdependent and interrelated. By taking a One Health approach, we can ensure that we are addressing health issues in a way that considers the full scope of their impact.

It is important to note that the One Health approach should not be anthropocentric. This means that we should not prioritize human interests above those of animals and the environment. Instead, we should ensure that we are taking a balanced approach that considers the welfare and interests of all three components equally.

In addition to promoting balance, the One Health approach should also maintain equity, ethics, and justice. This means that everyone, regardless of their economic status or social background, should have access to quality healthcare. It also means that ethical principles should guide our decision-making, prioritizing the welfare and rights of animals and the environment. Finally, we should work to promote social justice by addressing health disparities and ensuring that vulnerable populations are not left behind.

Finally, the One Health approach should be culture-oriented. This means that we should recognize and respect the diverse cultural practices and beliefs that influence health and well-being. We should also involve local communities in the design and implementation of health interventions to ensure that they are culturally appropriate and effective. By doing so, we can ensure that we are addressing health issues in a way that is respectful, inclusive, and effective.

References

1. World Health Organization (WHO) (2020) The one health definition and principles developed by OHHLEP translations. https://cdn.who.int/media/docs/default-source/one-health/ohhlep/one-health-definition-and-principles-translations.pdf?sfvrsn=d85839dd_6&download=true. Accessed date 19 Feb 2024
2. Wear A (2008) Place, health, and disease: the airs, waters, places tradition in early modern England and North America. J Medieval and Early Modern Stud 38(3):443–465. https://doi.org/10.1215/10829636-2008-003
3. Paleari A, Beretta EP, Riva MA (2021) Giovanni Maria Lancisi (1654–1720) and the modern cardiovascular physiology. Adv Physiol Educ 45(1):154–159. https://doi.org/10.1152/ADVAN.00218.2020
4. Schultz M (2008) Rudolf Virchow. Emerg Infect Dis 14(9):1480. https://doi.org/10.3201/EID1409.086672
5. Evans BR, Leighton FA (2014) A history of one health. Rev Sci Tech Off Int Epiz 33(2):413–420. https://doi.org/10.20506/rst.33.2.2298
6. Dukes T (2024) Veterinary medicine: an illustrated history. The Canadian Veterinary J 37(2):116. Accessed 25 Jan [Online]. Available https://www.ncbi.nlm.nih.gov/pmc/articles/PMC1576620/
7. Pal M, Gebrezabiher W, Rahman MT (2014) The roles of veterinary, medical and environmental professionals to achieve one health. J Adv Vet Anim Res 1(4):148–155. https://doi.org/10.5455/javar.2014.a27
8. Cassidy A (2017) Humans, other animals and 'One Health' in the early twenty-first century. Animals and the Shaping of Modern Med 193–236. https://doi.org/10.1007/978-3-319-64337-3_6
9. Zinsstag J (2024) One health, animal ethics and animal welfare. [Online]. https://www.futurelearn.com/info/courses/one-health/0/steps/25458. Accessed 25 Jan 2024
10. Rabinowitz PM, Odofin L, Dein FJ (2008) From 'Us vs. Them' to 'Shared Risk': can animals help link environmental factors to human health? Eco Health 5(2):224–229. https://doi.org/10.1007/s10393-008-0170-4
11. Cumming DHM, Cumming GS (2015) One health: an ecological and conservation perspective. In: One health: the theory and practice of integrated health approach, pp 38–52. https://doi.org/10.1079/9781780643410.0038
12. Brown C (2010) Human-animal medicine: clinical approaches to Zoonoses, toxicants, and other shared health risks. Emerg Infect Dis 16(6):1050. https://doi.org/10.3201/EID1606.100367
13. Lederman Z, Magalhães-Sant M, Chuan Voo T (2021) Stamping out animal culling: from anthropocentrism to one health ethics. J Agric Environ Ethics 34(5):27. https://doi.org/10.1007/s10806-021-09868-x
14. Van Herten J, Buikstra S, Bovenkerk B, Stassen E (2020) Ethical decision-making in zoonotic disease control how do one health strategies function in the Netherlands? J Agric Environ Ethics 33:239–259. https://doi.org/10.1007/s10806-020-09828-x

15. Anthony R, De Paula VA (2022) One health animal disaster management: an ethics of care approach. J Appl Anim Welfare Sci 25(2):180–194. https://doi.org/10.1080/10888705.2022.2040360
16. Diller ER, Williamson L (2023) Supporting one health for pandemic prevention: the need for ethical innovation. J Bioeth Inq 20(3):345. https://doi.org/10.1007/S11673-023-10264-5
17. Nieuwland J, Meijboom FLB (2019) One health: how interdependence enriches veterinary ethics education 10(1):13. https://doi.org/10.3390/ani10010013
18. Johnson J, Degeling C (2019) Does one health require a novel ethical framework? J Med Ethics 45(4):239–243. https://doi.org/10.1136/MEDETHICS-2018-105043
19. Beever J, Morar N (2019) The epistemic and ethical onus of one health. Bioethics 33(1):185–194. https://doi.org/10.1111/BIOE.12522
20. Coghlan S, Coghlan BJ, Capon A, Singer P (2021) A bolder one health: expanding the moral circle to optimize health for all. One Health Outlook 3(1):1–4. https://doi.org/10.1186/S42522-021-00053-8
21. Zinsstag J, Schelling E, Waltner-Toews D, Tanner M (2011) From 'one medicine' to 'one health' and systemic approaches to health and well-being. Prev Vet Med 101:148–156. https://doi.org/10.1016/j.prevetmed.2010.07.003
22. Swanson I (2022) One health, community engagement, and praxis: shifting the health equity paradigm [Online]. Available https://escholarship.org/uc/item/0238m7f4. Accessed 29 Jan 2024
23. Cleaveland S et al (2017) One health contribution towards more effective and equitable approaches to health in low- and middle-income countries. Philosophical Trans Royal Soc B: Biol Sci 372(1725):20160168. https://doi.org/10.1098/RSTB.2016.0168
24. Adisasmito WB, Almuhairi S (2023) One health action for health security and equity 401(10376):530–533. https://doi.org/10.1016/S0140-6736(23)00086-7
25. Morgan BL, Stern MC, Pérez-Stable EJ, Hooper MW, Fejerman L (2022) Adding a one health approach to a research framework for minority health and health disparities. Elife 11. https://doi.org/10.7554/ELIFE.76461
26. Morgan BL, Stern MC, Pérez-Stable EJ, Hooper MW, Fejerman L (2022) Adding a one health approach to a research framework for minority health and health disparities. Elife 11:76461. https://doi.org/10.7554/ELIFE.76461
27. Eliakimu ES, Mans L (2022) Addressing inequalities toward inclusive governance for achieving one health: a rapid review. Front Public Health 9:755285. https://doi.org/10.3389/FPUBH.2021.755285/BIBTEX
28. Solis A, Nunn CL (2021) One health disparities and COVID-19. Evol Med Public Health 9(1):70–77. https://doi.org/10.1093/mph/eoab003
29. Lysaght T et al. (2017) Justice is the missing link in one health: results of a mixed methods study in an urban City State. PLoS One 12(1). https://doi.org/10.1371/JOURNAL.PONE.0170967
30. Riadhussyah M, Jannah AW, Basir SM, Subarkah AR (2023) Forming international instrument through one health approach for health justice. J IUS Kajian Hukum dan Keadilan 11(3):493–511. https://doi.org/10.29303/ius.v11i3.1246
31. Hawes SM, Hupe T, Morris KN (2020) Punishment to support: the need to align animal control enforcement with the human social justice movement. Animals (Basel) 10(10):1–8. https://doi.org/10.3390/ANI10101902
32. Murray MH et al. (2022) One health for all: advancing human and ecosystem health in cities by integrating an environmental justice lens. https://doi.org/10.1146/annurev-ecolsys-102220
33. Malhotra KC, Gokhale Y, Chatterjee S (2001) Cultural and ecological dimensions of sacred groves in India. Indian National Science Academy. https://sacredland.org/wp-content/uploads/2017/07/malhotra-et-al.pdf

34. Nair S (2021) The Indian strategy of one health concept. J Alternat Complement Integrat Med 7(3):1–4. https://doi.org/10.24966/ACIM-7562/100183
35. Weiss MG, Agashe M, Gupte MD (2021) The concept of one health: cultural context, background and prospects in India. Indian J Med Res 153(3):333. https://doi.org/10.4103/IJMR.IJMR_732_21